Concrete Construction Handbook

Concrete Construction Handbook

Mark Chapman

STATES
ACADEMIC PRESS
www.statesacademicpress.com

States Academic Press,
109 South 5th Street,
Brooklyn, NY 11249, USA

ISBN: 978-1-63989-124-5

Cataloging-in-Publication Data

Concrete construction handbook / Mark Chapman.
p. cm.
Includes bibliographical references and index.
ISBN 978-1-63989-124-5
1. Concrete construction. 2. Concrete. 3. Civil engineering. I. Chapman, Mark.
TA681 .C66 2022
624.183 4--dc23

For information on all States Academic Press publications visit our website at www.statesacademicpress.com

STATES
ACADEMIC PRESS

Contents

Preface

This book has been an outcome of determined endeavour from a group of educationists in the field. The primary objective was to involve a broad spectrum of professionals from diverse cultural background involved in the field for developing new researches. The book not only targets students but also scholars pursuing higher research for further enhancement of the theoretical and practical applications of the subject.

A construction material composed of cement, fine aggregates such as sand and coarse aggregates which when mixed with water hardens with time is referred to as concrete. The study of properties of the concrete and its practical applications is referred to as concrete technology. During the construction of a building, concrete is used for the construction of foundations, columns, beams, slabs, and other load-bearing structures. Concrete is mixed and prepared in proportions with respect to cement quantity. The two types of concrete mixes are nominal mix and design mix. The nominal mix is used for normal construction works such as small residential buildings whereas for design mix concrete, the mix proportions are finalized based on various lab tests on cube or cylinder for its compressive strength. This book covers in detail some existent theories and innovative concepts revolving around concrete construction. It presents the complex subject of concrete construction in the most comprehensible and easy to understand language. This book is appropriate for students seeking detailed information in this area as well as for experts.

It was an honour to edit such a profound book and also a challenging task to compile and examine all the relevant data for accuracy and originality. I wish to acknowledge the efforts of the contributors for submitting such brilliant and diverse chapters in the field and for endlessly working for the completion of the book. Last, but not the least; I thank my family for being a constant source of support in all my research endeavours.

Mark Chapman

1

CONCRETE MATERIALS: COSTING, SELECTING AND PREPARING

Concrete is an artificial stone produced by mixing cement mortar with broken stone, gravel, broken slag, cinders or other similar fragmentary materials. The component parts are therefore hydraulic cement, sand and the broken stone or other coarse material commonly designated as the aggregate.

CEMENT.

At least a score of varieties of hydraulic cement are listed in the classifications of cement technologists. The constructing engineer and contractor recognize only three varieties: Portland cement, natural cement and slag or puzzolan cement. All concrete used in engineering work is made of either Portland, natural or slag cement, and the great bulk of all concrete is made of Portland cement. Only these three varieties of cement are, therefore, considered here and they only in their aspects having relation to the economics of construction work. For a full discussion of the chemical and physical properties of hydraulic cements and for the methods of determining these properties by tests, the reader is referred to "Practical Cement Testing," by W. Purves Taylor.

PORTLAND CEMENT.—Portland cement is the best of the hydraulic cements. Being made from a rigidly controlled artificial mixture of lime, silica and alumina the product of the best mills is a remarkably strong, uniform and stable material. It is suitable for all classes of concrete work and is the only variety of hydraulic cement allowable for reinforced concrete or for plain concrete having to endure hard wear or to be used where strength, density and durability of high degree are demanded.

NATURAL CEMENT.—Natural cement differs from Portland cement in degree only. It is made by calcining and grinding a limestone rock containing naturally enough clayey matter (silica and alumina) to make a cement that will harden under water. Owing to the imperfection and irregularity of the natural rock mixture, natural cement is weaker and less uniform than Portland cement. Natural cement concrete is suitable for work in which great unit strength or uniformity of quality is not essential. It is never used for reinforced work.

SLAG CEMENT.—Slag cement has a strength approaching very closely that of Portland cement, but as it will not stand exposure to the air slag cement concrete is suitable for use only under water. Slag cement is made by grinding together slaked lime and granulated blast furnace slag.

SIZE AND WEIGHT OF BARRELS OF CEMENT.—The commercial unit of measurement of cement is the barrel; the unit of shipment is the bag. A barrel of Portland cement contains 380 lbs. of cement, and the barrel itself weighs 20 lbs.; there are four bags (cloth or paper sacks) of cement to the barrel, and the regulation cloth sack weighs 1½ lbs. The size of cement barrels varies, due to the differences in weight of cement and to differences in compacting the cement into the barrel. A light burned Portland cement weighs 100 lbs. per struck bushel; a heavy burned Portland cement weighs 118 to 125 lbs. per struck bushel. The number of cubic feet of packed Portland cement in a barrel ranges from 3 to 3½. Natural cements are lighter than Portland cement. A barrel of Louisville, Akron, Utica or other Western natural cement contains 265 lbs. of cement and weighs 15 lbs. itself; a barrel of Rosendale or other Eastern cement contains 300 lbs. of cement and the barrel itself weighs 20 lbs. There are 3¾ cu. ft. in a barrel of Louisville cement. Usually there are three bags to a barrel of natural cement.

As stated above, the usual shipping unit for cement is the bag, but cement is often bought in barrels or, for large works, in bulk. When bought in cloth bags, a charge is made of 10 cts. each for the bags, but on return of the bags a credit of 8 to 10 cts. each is allowed. Cement bought in barrels costs 10 cts. more per barrel than in bulk, and cement ordered in paper bags costs 5 cts. more per barrel than in bulk. Cement is usually bought in cloth sacks which are returned, but to get the advantage of this method of purchase the user must have an accurate system for preserving, checking up and shipping the bags.

Where any considerable amount of cement is to be used the contractor will find that it will pay to erect a small bag house or to close off a room at the mixing plant. Provide the enclosure with a locked door and with a small window into which the bags are required to be thrown as fast as emptied. One trustworthy man is given the key and the task of counting up the empty bags each day to see that they check with the bags of cement used. The following rule for packing and shipping is given by Gilbreth.[A]

[A] "Field System," Frank B. Gilbreth. Myron C. Clark Publishing Co., New York and Chicago.

"Pack cement bags laid flat, one on top of the other, in piles of 50. They can then be counted easily. Freight must be prepaid when cement bags are returned and bills of lading must be obtained in duplicate or credit cannot be obtained on shipment."

The volumes given above are for cement compacted in the barrel. When the cement is emptied and shoveled into boxes it measures from 20 to 30 per cent more than when packed in the barrel. The following table compiled from tests made for the Boston Transit Commission, Mr. Howard Carson, Chief Engineer, in 1896, shows the variation in volume of cement measured loose and packed in barrels:

Brand	Vol. Barrel cu. ft.	Vol. Packed cu. ft.	Vol. Loose cu. ft.	Per cent Increase in bulk
Portland.				
Giant	3.5	3.35	4.17	25
Atlas	3.45	3.21	3.75	18
Saylors	3.25	3.15	4.05	30
Alsen	3.22	3.16	4.19	33
Dyckerhoff	3.12	3.03	4.00	33

Mr. Clarence M. Foster is authority for the statement that Utica cement barrels measure 16¼ ins. across at the heads, 19½ ins. across the bilge, and 25¾ ins. in length under heads, and contain 3.77 cu. ft. When 265 lbs. of Utica natural hydraulic cement are packed in a barrel it fills it within 2½ ins. of the top and occupies 3.45 cu. ft., and this is therefore the volume of a barrel of Utica hydraulic cement packed tight.

In comparative tests made of the weights and volumes of various brands of cements at Chicago in 1903, the following figures were secured:

	Vol. per bbl., cu. ft.	Weight per bbl., lbs.		Weight per cu. ft.
Brand.	Loose.	Gross.	Net.	Loose, lbs.
Dyckerhoff	4.47	395	369.5	83
Atlas	4.45	401	381	85.5
Alpha	4.37	400.5	381	86.5
Puzzolan	4.84	375	353.5	73.5
Steel	4.96	345	322.5	67.5
Hilton	4.64	393	370.5	79.5

SPECIFICATIONS AND TESTING—The great bulk of cement used in construction work is bought on specification. The various government bureaus, state and city works departments, railway companies, and most public service corporations have their own specifications. Standard specifications are also put forward by several of the national engineering societies, and one of these or the personal specification of the engineer is used for individual works. Buying cement to specification necessitates testing to determine that the material purchased meets the specified requirements. For a complete discussion of the methods of conducting such tests the reader is referred to "Practical Cement Testing" by W. Purves Taylor.

According to this authority a field testing laboratory will cost for equipment $250 to $350. Such a laboratory can be operated by two or three men at a salary charge of from $100 to $200 per month. Two men will test on an average four samples per day and each additional man will test four more samples. The cost of testing will range from $3 to $5 per sample, which is roughly equivalent to 3 cts. per barrel of cement, or from 3 to 5 cts. per cubic yard of concrete. These figures are for field laboratory work reasonably well conducted under ordinarily favorable conditions. In large laboratories the cost per sample will run somewhat lower.

SAND.

Sand constitutes from ⅓ to ½ of the volume of concrete; when a large amount of concrete is to be made a contractor cannot, therefore, afford to guess at his source of sand supply. A long haul over poor roads can easily make the sand cost more than the stone per cubic yard of concrete.

PROPERTIES OF GOOD SAND.—Engineers commonly specify that sand for concrete shall be clean and sharp, and silicious in character. Neither sharpness nor excessive cleanliness is worth seeking after if it involves much expense. Tests show conclusively that sand with rounded grains makes quite as strong a mortar, other things being equal, as does sand with angular grains. The admixture with sand of a considerable percentage of loam or clay is also not the unmixed evil it has been supposed to be. Myron S. Falk records[B] a number of elaborate experiments on this point. These experiments demonstrate conclusively that loam and clay in sand to the amount of 10 to 15 per cent. result in no material reduction in the strength of mortars made with this sand as compared with mortars made with the same sand after washing. There can be no doubt but that for much concrete work the expense entailed in washing sand is an unnecessary one.

[B] "Cements, Mortars and Concretes" By Myron S. Falk. Myron C. Clark Publishing Co., Chicago, Ill.

The only substitute for natural sand for concrete, that need be considered practically, is pulverized stone, either the dust and fine screenings produced in crushing rock or an artificial sand made by reducing suitable rocks to powder. As a conclusion from the records of numerous tests, M. S. Falk says: "It may be concluded that rock screenings may be substituted for sand, either in mortar or concrete, without any loss of strength resulting. This is important commercially, for it precludes the necessity of screening the dust from crushed rock and avoids, at the same time, the cost of procuring a natural sand to take its place."

The principal danger in using stone dust is failure to secure the proper balance of different size grains. This is also an important matter in the choice of natural sands. Sand composed of a mixture of grains ranging from fine to coarse gives uniformly stronger mortars than does sand with grains of nearly one size, and as between a coarse and a fine sand of one size of grains the coarse sand gives the stronger mortar. Further data on the effect of size of grains on the utility of sand for concrete are given in Chapter II, in the section on Voids in Sand, and for those who wish to study in detail, the test data on this and the other matters referred to here, the authors recommend

"Cements, Mortars and Concretes; Their Physical Properties," by Myron S. Falk.

COST OF SAND.—A very common price for sand in cities is $1 per cu. yd., delivered at the work. It may be noted here that as sand is often sold by the load instead of the cubic yard, it is wise to have a written agreement defining the size of a load. Where the contractor gets his sand from the pit its cost will be the cost of excavating and loading at the pit, the cost of hauling in wagons, the cost of freight and rehandling it if necessary, and the cost of washing, added together.

An energetic man working under a good foreman will load 20 cu. yds. of sand into wagons per 10-hour day; with a poor foreman or when laborers are scarce, it is not safe to count on more than 15 cu. yds. per day. With wages at $1.50 per day this will make the cost of loading 10 cts. per cubic yard. The cost of hauling will include the cost of lost team time and dumping, which will average about 5 cts. per cubic yard. With 1 cu. yd. loads, wages of team 35 cts. per hour, and speed of travel 2½ miles per hour, the cost of hauling proper is ½ ct. per 100 ft., or 27 cts. per mile. Assuming a mile haul, the cost of sand delivered based on the above figures will be 10 cts. + 5 cts. + ½ ct. per 100 ft. = 15 + 27 cts. = 42 cts. per cu. yd. Freight rates can always be secured and it is usually safe to estimate the weight on a basis of 2,700 lbs. per cubic yard. For a full discussion of the cost of excavating sand and other earths the reader is referred to "Earth Excavation and Embankments; Methods and Cost," by Halbert P. Gillette and Daniel J. Hauer.

METHODS AND COST OF WASHING SAND.—When the available sand carries considerable percentages of loam or clay and the specifications require that clean sand shall be used, washing is necessary. The best and cheapest method of performing this task will depend upon the local conditions and the amount of sand to be washed.

Washing With Hose.—When the quantity of sand to be washed does not exceed 15 to 30 cu. yds. per day the simplest method, perhaps, is to use a hose. Build a wooden tank or box, 8 ft. wide and 15 ft. long, the bottom having a slope of 8 ins. in the 15 ft. The sides should be about 8 ins. high at the lower end and rise gradually to 3 ft. in height at the upper end. Close the lower end of the tank with a board gate about 6 ins. in height and sliding in grooves so that it can be removed. Dump about 3 cu. yds. of sand into the upper end of the tank and play a ¾-in. hose stream of water on it, the hose man standing at the lower end of the tank. The water and sand flow down the inclined bottom of the tank where the sand remains and the dirt flows over the gate and off with the water. It takes about an hour to wash a 3-cu. yd. batch, and by building a pair of tanks so that the hose man can shift from one to the other, washing can proceed continuously and one man will wash 30 cu. yds. per 10-hour day at a cost, with wages at $1.50, of 5 cts. per cubic yard. The sand, of course, has to be shoveled from the tank and this will cost about 10 cts. per cubic yard, making 15 cts. per cubic yard for washing and shoveling, and to this must be added any extra hauling and, if the water is pumped, the cost of pumping which may amount to 10 cts. per cubic yard for coal and wages. Altogether a cost of from 15 to 30 cts. per cubic yard may be figured for washing sand with a hose.

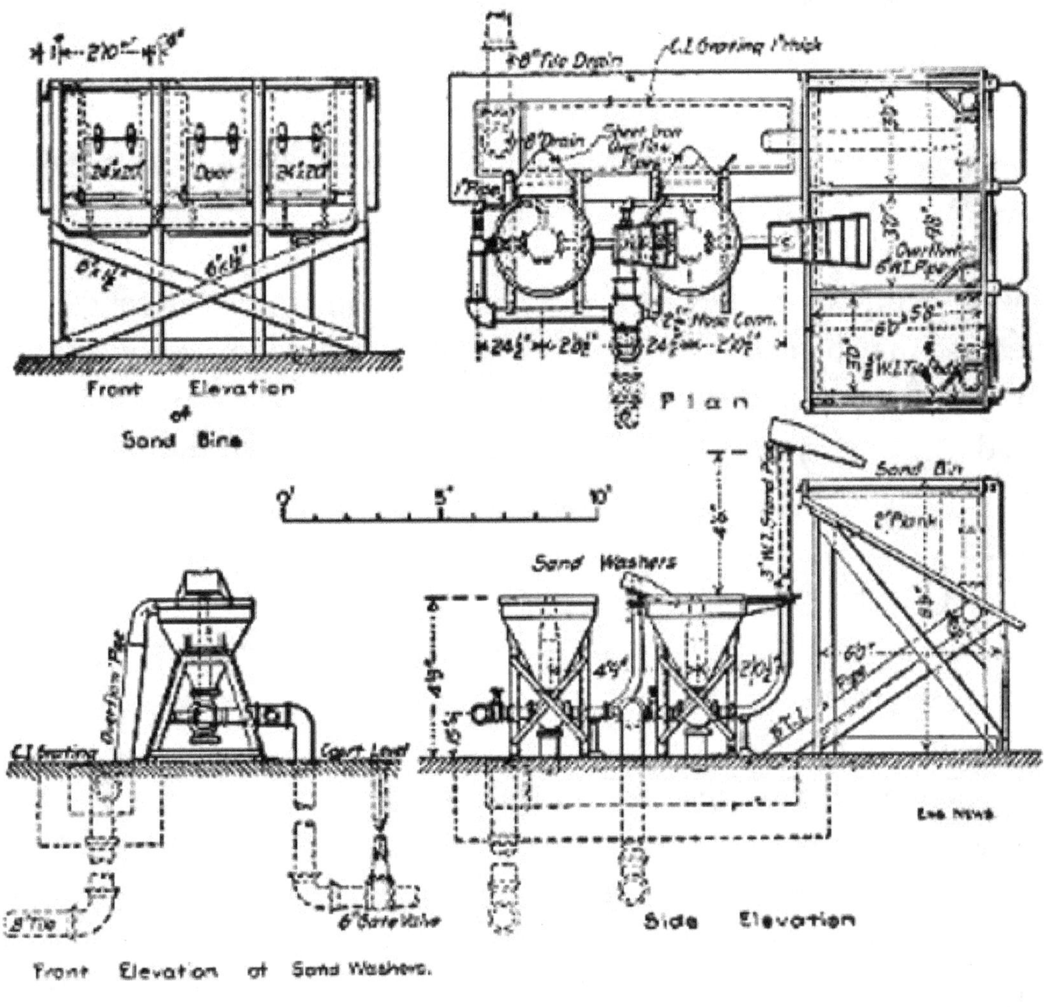

Fig. 1.—Plan and Elevation of Two-Hopper Ejector Sand Washing Plant.

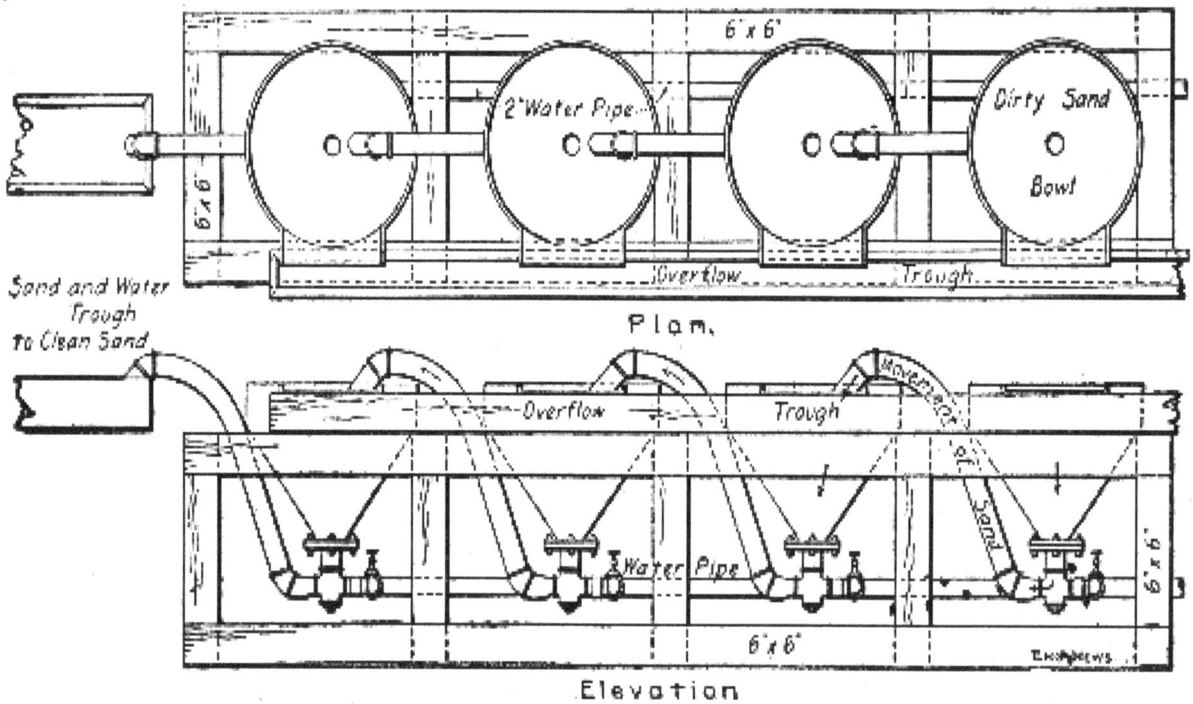

Fig. 2.—Plan and Elevation of Four-Hopper Ejector Sand Washing-Plant.

Washing With Sand Ejectors.—When large quantities of sand are to be washed use may be made of the sand ejector system, commonly employed in washing filter sand at large water filtration plants; water under pressure is required. In this system the dirty sand is delivered into a conical or pyramidal hopper, from the bottom of which it is drawn by an ejector and delivered mixed with water into a second similar hopper; here the water and dirt overflow the top of the hopper, while the sand settles and is again ejected into a third hopper or to the stock pile or bins. The system may consist of anywhere from two to six hoppers. Figure 1 shows a two-hopper lay-out and Fig. 2 shows a four-hopper lay-out. In the first plant the washed sand is delivered into bins so arranged, as will be seen, that the bins are virtually a third washing hopper. The clean sand is chuted from these bins directly into cars or wagons. In the second plant the clean sand is ejected into a trough which leads it into buckets handled by a derrick. The details of one of the washing hoppers for the plant shown by Fig. 1 are illustrated by Fig. 3.

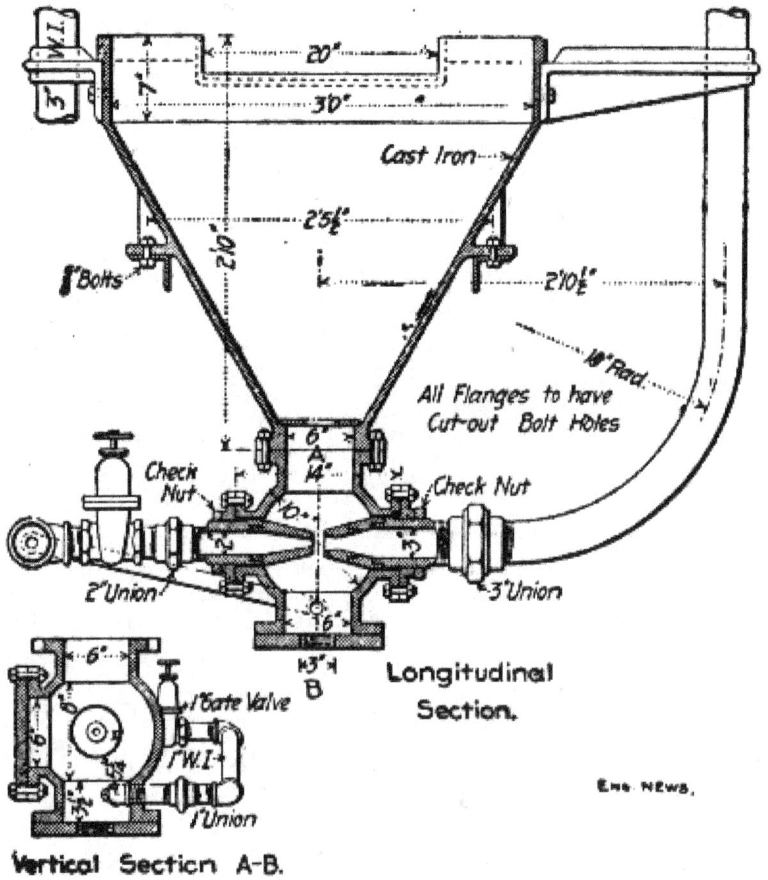

Fig. 3.—Details of Washing Hopper and Ejector for Plant Shown by Fig. 1.

At filter plants the dirty sand is delivered mixed with water to the first hopper by means of ejectors stationed in the filters and discharging through pipes to the washers. When, as would usually be the case in contract work, the sand is delivered comparatively dry to the first hopper, this hopper must be provided with a sprinkler pipe to wet the sand. In studying the ejector washing plants illustrated it should be borne in mind that for concrete work they would not need to be of such permanent construction as for filter plants, the washers would be mounted on timber frames, underground piping would be done away with, etc.; at best, however, such plants are expensive and will be warranted only when the amount of sand to be washed is large.

The usual assumption of water-works engineers is that the volume of water required for washing filter sand is 15 times the volume of the sand washed. At the Albany, N. Y., filters the sand passes through five ejectors at the rate of 3 to 5 cu. yds. per hour and takes 4,000 gallons of water per cubic yard. One man shovels sand into the washer and two take it away. Based on an output of 32 cu. yds. in 10 hours, Mr. Allen Hazen estimates the cost of washing as follows:

3 men, at $2 per day	$6.00
110,000 gallons of water, at $0.05	5.50
Total, 32 cu. yds., at 36 cts.	$11.50

Washing With Tank Washers.—Figure 4 shows a sand washer used in constructing a concrete lock at Springdale, Pa., in the United States government improvement work on the Allegheny river. The device consisted of a circular tank 9 ft. in diameter and 7 ft. high, provided with a sloping false bottom perforated with 1-in. holes, through which water was forced as indicated. A 7½×5×6-in. pump with a 3-in. discharge pipe was used to force water into the tank, and the rotating paddles were operated by a 7 h.p. engine. This apparatus washed a batch of 14 cu. yds. in from 1 to 2 hours at a cost of 7 cts. per cubic yard. The sand contained much fine coal and silt. The above data are given by Mr. W. H. Roper.

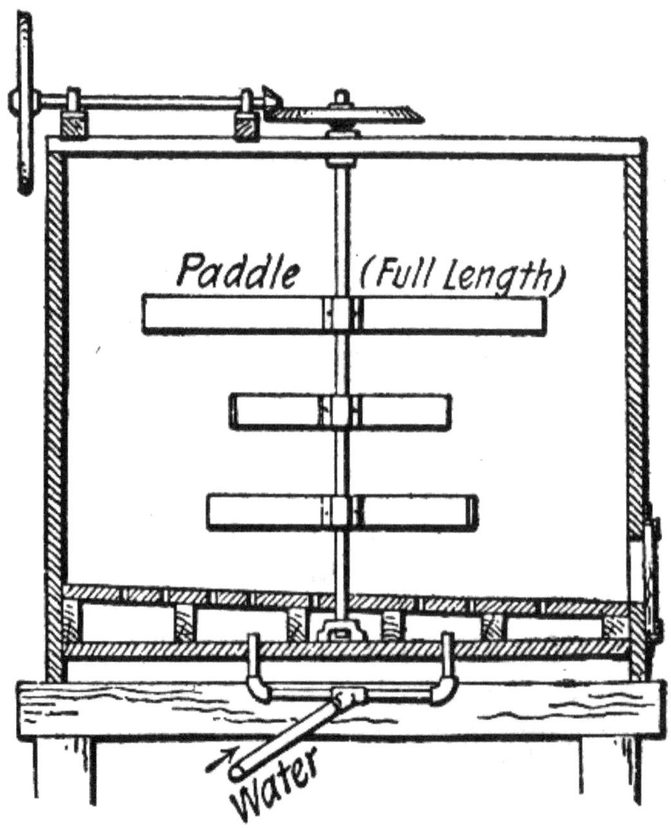

Fig. 4.—Details of Tank Washer Used at Springdale, Pa.

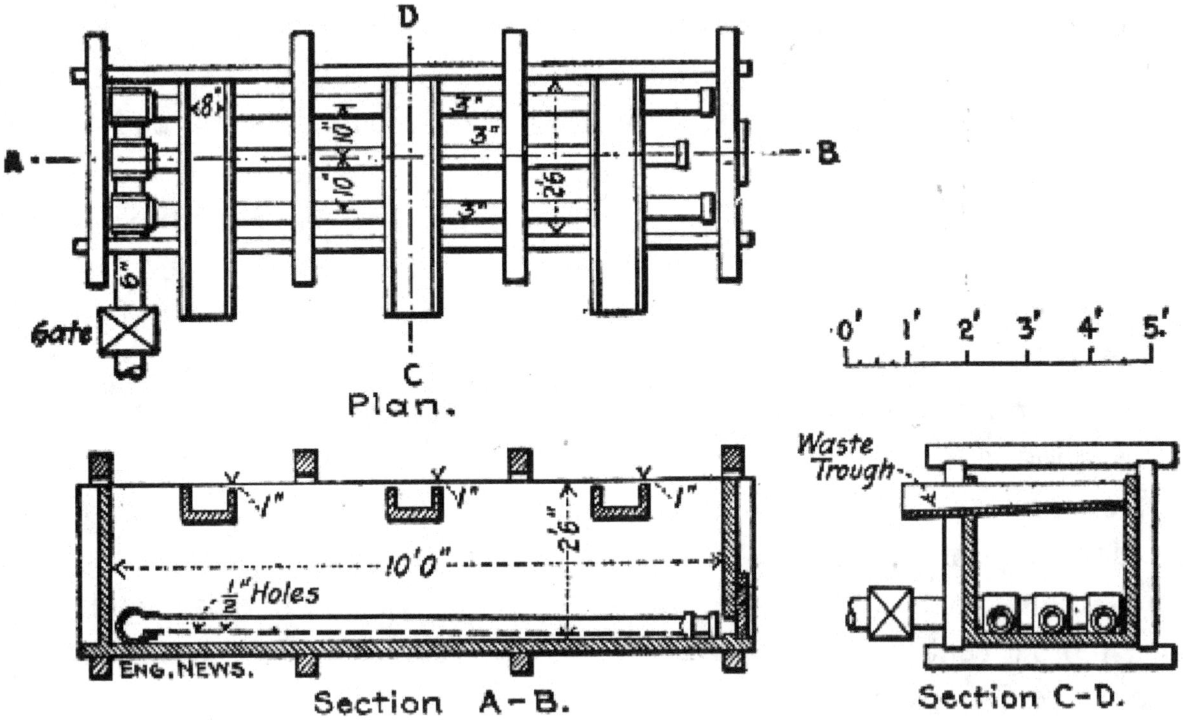

Fig. 5.—Details of Tank Washer Used at Yonkers, N. Y.

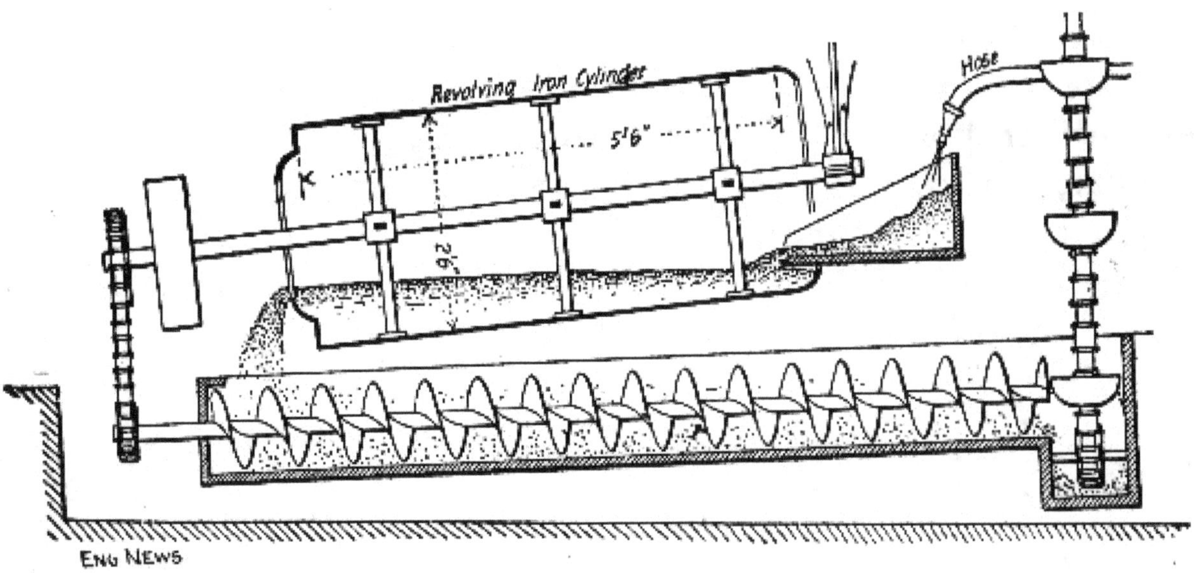

Fig. 6.—Details of Rotating Tank Sand Washer Used at Hudson, N. Y.

Another form of tank washer, designed by Mr. Allen Hazen, for washing bank sand at Yonkers, N. Y., is shown by Fig. 5. This apparatus consisted of a 10×2½×2½ ft. wooden box, with a 6-in. pipe entering one end at the bottom and there branching into three 3-in. pipes, extending along the bottom and capped at the ends. The undersides of the 3-in. pipes were pierced with ½-in. holes 6 ins. apart, through which water under pressure was discharged into the box. Sand was shoveled into the box at one end and the upward currents of water raised the fine and dirty particles until they escaped through the waste troughs. When the box became filled with sand a sliding door at one end was opened and the batch discharged. The operation was continuous as long as sand was shoveled into the box; by manipulating the door the sand could be made to run out with a very small percentage of water. Sand containing 7 per cent of dirt was thus washed so that it contained only 0.6 per cent dirt. The washer handled 200 cu. yds. of sand in 10 hours. The above data are given by F. H. Stephenson.

A somewhat more elaborate form of tank washer than either of those described is shown by Fig. 6. This apparatus was used by Mr. Geo. A. Soper for washing filter sand at Hudson, N. Y. The dirty sand was shoveled into a sort of hopper, from which it was fed by a hose stream into an inclined cylinder, along which it traveled and was discharged into a wooden trough provided with a screw conveyor and closed at both ends. The water overflowing the sides of the trough carried away the dirt and the clean sand was delivered by the screw to the bucket elevator which hoisted it to a platform, from which it was taken by barrows to the stock pile. A 4-h.p. engine with a 5-h.p. boiler operated the cylinder, screw, elevator and pump. Four men operated the washer and handled 32 cu. yds. of sand per day; with wages at $1.50 the cost of washing was 20 cts. per cubic yard.

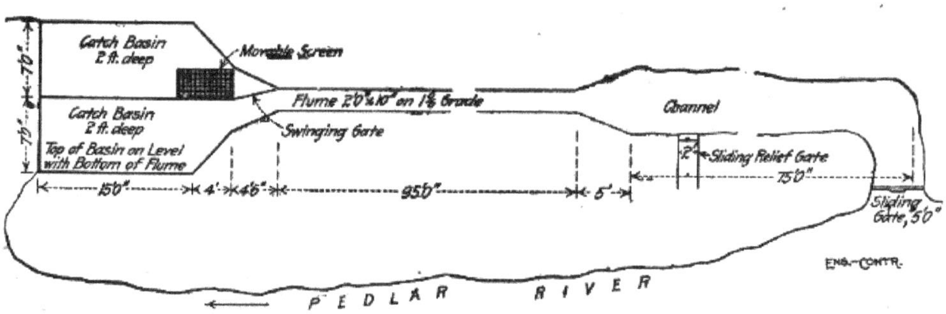

Fig. 7.—Arrangement of Sand Washing Plant at Lynchburg, Va.

In constructing a concrete block dam at Lynchburg, Va., sand containing from 15 to 30 per cent. of loam, clay and vegetable matter was washed to a cleanliness of 2 to 5 per cent of such matter by the device shown by Fig. 7. A small creek was diverted, as shown, into a wooden flume terminating in two sand tanks; by means of the swinging gate the flow was passed through either tank as desired. The sand was hauled by wagon and shoveled into the upper end of the flume; the current carried it down into one of the tanks washing the dirt loose and carrying it off with the overflow over the end of the tank while the sand settled in the tank. When one tank was full the flow was diverted into the other tank and the sand in the first tank was shoveled out, loaded into wagons, and hauled to the stock pile. As built this washer handled about 30 cu. yds. of sand per 10-hour day, but the tanks were built too small for the flume, which could readily handle 75 cu. yds. per day with no larger working force. This force consisted of three men at $1.50 per day, making the cost, for a 30 cu. yd. output, 15 cts. per cu. yd. for washing.

None of the figures given above includes the cost of handling the sand to and from the washer. When this involves much extra loading and hauling, it amounts to a considerable expense, and in any plan for washing sand the contractor should figure, with exceeding care, the extra handling due to the necessity of washing.

AGGREGATES.

The aggregates commonly used in making concrete are broken or crushed stone, gravel, slag and cinders. Slag and cinders make a concrete that weighs considerably less than stone or gravel mixtures, and being the products of combustion are commonly supposed to make a specially fire resisting concrete; their use is, therefore, confined very closely to fireproof building work and, in fact, to floor construction for such buildings. Slag and cinder concretes are for this reason given minor consideration in this volume.

BROKEN STONE.—Stone produced by crushing any of the harder and tougher varieties of rock is suitable for concrete. Perhaps the best stone is produced by crushing trap rock. Crushed trap besides being hard and tough is angular and has an excellent fracture surface for holding cement; it also withstands heat better than most stone. Next to trap the hard, tough, crystalline limestones make perhaps the best all around concrete material; cement adheres to limestone better than to any other rock. Limestone, however, calcines when subjected to fire and is, therefore, objected to by many engineers for building construction. The harder and denser sandstones, mica-schists, granites and syanites make good stone for concrete and occasionally shale and slate may be used.

GRAVEL.—Gravel makes one of the best possible aggregates for concrete. The conditions under which gravel is produced by nature make it reasonably certain that only the tougher and harder rocks enter into its composition; the rounded shapes of the component particles permit gravel to be more closely tamped than broken stone and give less danger of voids from bridging; the mixture is also generally a fairly well balanced composition of fine and coarse particles. The surfaces of the particles being generally smooth give perhaps a poorer bond with the cement than most broken stone. In the matter of strength the most recent tests show that there is very little choice between gravel and broken stone concrete.

SLAG AND CINDERS.—The slag used for concrete aggregate is iron blast furnace slag crushed to proper size. Cinders for aggregate are steam boiler cinders; they are best with the fine ashes screened out and should not contain more than 15 per cent. of unburned coal.

BALANCED AGGREGATE.—With the aggregate, as with the sand for concrete, the best results, other things being equal, will be secured by using a well-balanced mixture of coarse and fine particles. Usually the product of a rock crusher is fairly well balanced except for the very fine material. There is nearly always a deficiency of this, which, as explained in a succeeding section, has to be supplied by adding sand. Usually, also, the engineer accepts the crusher product coarser than screenings as being well enough balanced for concrete work, but this is not always the case. Engineers occasionally demand an artificial mixture of varying proportions of different size stones and may even go so far as to require gravel to be screened and reproportioned. This artificial grading of the aggregate adds to the cost of the concrete in some proportion which must be determined for each individual case.

SIZE OF AGGREGATE.—The size of aggregate to be used depends upon the massiveness of the structure, its purpose, and whether or not it is reinforced. It is seldom that aggregate larger than will pass a 3-in. ring is used and this only in very massive work. The more usual size is 2½ ins. For reinforced concrete 1¼ ins. is about the maximum size allowed and in building work 1-in. aggregate is most commonly used. Same constructors use no aggregate larger than ¾ in. in reinforced building work, and others require that for that portion of the concrete coming directly in contact with the reinforcement the aggregate shall not exceed ¼ to ½ in. The great bulk of concrete work is done with aggregate smaller than 2 ins., and as a general thing where the massiveness of the structure will allow of much larger sizes it will be more economic to use rubble concrete. (See Chapter VI.)

COST OF AGGREGATE.—The locality in which the work is done determines the cost of the aggregate. Concerns producing broken stone or screened and washed gravel for concrete are to be found within shipping distance in most sections of the country so that these materials may be purchased in any amount desired. The cost will then be the market price of the material f. o. b. cars at plant plus the freight rates and the cost of unloading and haulage to the stock piles. If the contractor uses a local stone or gravel the aggregate cost will be, for stone the costs of quarrying and crushing and transportation, and, for gravel, the cost of excavation, screening, washing and transportation.

SCREENED OR CRUSHER-RUN STONE FOR CONCRETE.—Formerly engineers almost universally demanded that broken stone for concrete should have all the finer particles screened out. This practice has been modified to some considerable extent in recent years by using all the crusher product both coarse and fine, or, as it is commonly expressed, by using run-of-crusher stone. The comparative merits of screened and crusher-run stone for concrete work are questions of comparative economy and convenience. The fine stone dust and chips produced in crushing stone are not, as was once thought, deleterious; they simply take the place of so much of the sand which would, were the stone screened, be required to balance the sand and stone mixture. It is seldom that the proportion of chips and dust produced in crushing stone is large enough to replace the sand constituent entirely; some sand has nearly always to be added to run-of-crusher stone and it is in determining the amount of this addition that uncertainty lies. The proportions of dust and chips in crushed stone vary with the kind of stone and with the kind of crusher used. Furthermore, when run-of-crusher stone is chuted from the crusher into a bin or pile the screenings and the coarse stones segregate. Examination of a crusher-run stone pile will show a cone-shaped heart of fine material enclosed by a shell of coarser stone, consequently when this pile of stone is taken from to make concrete a uniform mixture of fine and coarse particles is not secured, the material taken from the outside of the pile will be mostly coarse and that from the inside mostly fine. This segregation combined with the natural variation in the crusher product makes the task of adding sand and producing a balanced sand and stone mixture one of extreme uncertainty and some difficulty unless considerable expenditure is made in testing and reproportioning. When the product of the crusher is screened the task of proportioning the sand to the stone is a straightforward operation, and the screened out chips and dust can be used as a portion of the sand if desired. The only saving, then, in using crusher-run stone direct is the very small one of not having to screen out the fine material. The conclusion must be that the economy of unscreened stone for concrete is a very doubtful quantity, and that the risk of irregularity in unscreened stone mixtures is a serious one. The engineer's specifications will generally determine for the contractor whether he is to use screened or crusher-run stone, but these same specifications will not guarantee the regularity of the resulting concrete mixture; this will be the contractor's burden and if the engineer's inspection is rigid and the crusher-run product runs uneven for the reasons given above it will be a burden of considerable expense. The contractor will do well to know his product or to know his man before bidding less or even as little on crusher-run as on screened stone concrete.

COST OF QUARRYING AND CRUSHING STONE.—The following examples of the cost of quarrying and crushing stone are fairly representative of the conditions which would prevail on ordinary contract work. In quarrying and crushing New Jersey trap rock with gyratory crushers the following was the cost of producing 200 cu. yds. per day:

	Per day.	Per cu. yd.
3 drillers at $2.75	$ 8.25	$0.041
3 helpers at $1.75	5.25	0.026
10 men barring out and sledging	15.00	0.075
14 men loading carts	21.00	0.105
4 cart horses	6.00	0.030
2 cart drivers	3.00	0.015
2 men dumping carts and feeding crusher	3.00	0.015
1 fireman for drill boiler	2.50	0.013
1 engineman for crusher	3.00	0.015
1 blacksmith	3.00	0.015
1 blacksmith helper	2.00	0.010
1 foreman	5.00	0.025
2 tons coal at $3.50	7.00	0.035
150 lbs. 40% dynamite at 15 cts.	22.50	0.113
Total	$106.50	$0.533

The quarry face worked was 12 to 18 ft., and the stone was crushed to 2-in. size. Owing to the seamy character of the rock it was broken by blasting into comparatively small pieces requiring very little sledging. The stone was loaded into one-horse dump carts, the driver taking one cart to the crusher while the other was being loaded. The haul was 100 ft. The carts were dumped into an inclined chute leading to a No. 5 Gates crusher. The stone was elevated by a bucket elevator and screened. All stone larger than 2 ins. was returned through a chute to a No. 3 Gates crusher for recrushing. The cost given above does not include interest, depreciation, and repairs; these items would add about $8 to $10 more per day or 4 to 5 cts. per cubic yard.

In quarrying limestone, where the face of the quarry was only 5 to 6 ft. high, and where the amount of stripping was small, one steam drill was used. This drill received its steam from the same boiler that supplied the crusher engine. The drill averaged 60 ft. of hole drilled per 10-hr. day, but was poorly handled and frequently laid off for repairs. The cost of quarrying and crushing was as follows:

Quarry.

1 driller	$ 2.50
1 helper	1.50
1 man stripping	1.50
4 men quarrying	6.00
1 blacksmith	2.50
⅛ ton coal at $3	1.00
Repairs to drill	.60
Hose, drill steel and interest on plant	.90
24 lbs. dynamite	3.60
Total	$20.10

Crusher.

1 engineman	$ 2.50
2 men feeding crusher	3.50
6 men wheeling	9.00
1 bin man	1.50
1 general foreman	3.00
⅓ ton coal at $3	1.00
1 gallon oil	.25
Repairs to crusher	1.00
Repairs to engine and boiler	1.00
Interest on plant	1.00
Total	$23.75

Summary:

	Per day.	Per. cu. yd.
Quarrying	$20.10	$0.37
Crushing	23.75	0.39
Total for 60 cu. yds.	$43.85	$0.76

The "4 men quarrying" barred out and sledged the stone to sizes that would enter a 9×16-in. jaw crusher. The "6 men wheeling" delivered the stone in wheelbarrows to the crusher platform, the run plank being never longer than 150 ft. Two men fed the stone into the crusher, and a bin-man helped load the wagons from the bin, and kept tally of the loads. The stone was measured loose in the wagons, and it was found that the average load was 1½ cu. yds., weighing 2,400 lbs. per cu. yd. There were 40 wagon loads, or 60 cu. yds. crushed per 10-hr. day, although on some days as high as 75 cu. yds. were crushed. The stone was screened through a rotary screen, 9 ft. long, having three sizes of openings, ½-in., 1¼-in. and 2¼-in. The output was 16% of the smallest size, 24% of the middle size, and 60% of the large size. All tailings over 2½ ins. in size were recrushed.

It will be noticed that the interest on the plant is quite an important item. This is due to the fact that, year in and year out, a quarrying and crushing plant seldom averages more than 100 days actually worked per year, and the total charge for interest must be distributed over these 100 days, and not over 300 days as is so commonly and erroneously done. The cost of stripping the earth off the rock is often considerably in excess of the above given cost, and each case must be estimated separately. Quarry rental or royalty is usually not in excess of 5 cts. per cu. yd., and frequently much less. The dynamite used was 40%, and the cost of electric exploders is included in the cost given. Where a higher quarry face is used the cost of drilling and the cost of explosives per cu. yd. is less. Exclusive of quarry rent and heavy stripping costs, a contractor should be able to quarry and crush limestone or sandstone for not more than 75 cts. per cu. yd., or 62 cts. per ton of 2,000 lbs., wages and conditions being as above given.

The labor cost of erecting bins and installing a 9×16 jaw crusher, elevator, etc., averages about $75, including hauling the plant two or three miles, and dismantling the plant when work is finished.

The following is a record of the cost of crushing stone and cobbles on four jobs at Newton, Mass., in 1891. On jobs A and B the stone was quarried and crushed; on jobs C and D cobblestones were crushed. A 9×15-in. Farrel-Marsondon crusher was used, stone being fed in by two laborers. A rotary screen having ½, 1 and 2½-in. openings delivered the stone into bins having four compartments, the last receiving the "tailings" which had failed to pass through the screen. The broken stone was measured in carts as they left the bin, but several cart loads were weighed, giving the following weights per cubic foot of broken stone:

	Size.			
	½-in. lbs.	1-in. lbs.	2½-ins. lbs.	Tailings. lbs.
Greenish trap rock, "A"	95.8	84.3	88.3	91.0
Conglomerate, "B"	101.0	87.7	94.4	....
Cobblestones, "C" and "D"	102.5	98.0	99.6	....

A one-horse cart held 26 to 28 cu. ft. (average 1 cu. yd.) of broken stone; a two-horse cart, 40 to 42 cu. ft., at the crusher.

	Job.			
	A.	B.	C.	D.
Hours run	412	144	101	198
Short tons per hour	9.0	11.2	15.7	12.1
Cu. yds. per hour	7.7	8.9	11.8	9.0
Per cent of tailings	31.8	29.3	17.5	20.5
Per cent of 2½-in. stone	51.3	51.9	57.0	55.1
Per cent of 1-in. stone	10.2	....	....	....
Per cent of ½-in. stone or dust	6.7	18.8	25.5	23.4

Job.

	A.	B.	C.	D.
Explosives, coal for drill and repairs	$0.084	$0.018	....	....
Labor steam drilling	0.092	....	....	....
Labor hand drilling	....	0.249	....	....
Sharpening tools	0.069	0.023	....	....
Sledging stone for crusher	0.279	0.420	....	....
Loading carts	0.098	0.127	....	$0.144
Carting to crusher	0.072	0.062	$0.314	0.098
Feeding crusher	0.053	0.053	0.033	0.065
Engineer of crusher	0.031	0.038	0.029	0.036
Coal for crusher	0.079	0.050	0.047	0.044
Repairs to crusher	0.041	....	....	0.011
Moving portable crusher	....	0.023	....	0.019
Watchman ($1.75 a day)	....	0.053	0.022	0.030
Total cost per cu. yd.	$0.898	$1.116	$0.445	$0.447
Total cost per short ton	0.745	0.885	0.330	0.372

Note.—"A" was trap rock; "B" was conglomerate rock; "C" and "D" were trap and granite cobblestones. Common laborers on jobs "A" and "D" were paid $1.75 per 9-hr. day; on jobs "B" and "C," $1.50 per 9-hr. day; two-horse cart and driver, $5 per day; blacksmith, $2.50; engineer on crusher, $2 on job "A," $2.25 on "B," $2.00 on "C," $2.50 on "D"; steam driller received $3, and helper $1.75 a day; foreman, $3 a day. Coal was $5.25 per short ton. Forcite powder, 11⅓ cts. per lb.

For a full discussion of quarrying and crushing methods and costs and for descriptions of crushing machinery and plants the reader is referred to "Rock Excavation; Methods and Cost," by Halbert P. Gillette.

SCREENING AND WASHING GRAVEL.—Handwork is resorted to in screening gravel only when the amount to be screened is small and when it is simply required to separate the fine sand without sorting the coarser material into sizes. The gravel is shoveled against a portable inclined screen through which the sand drops while the pebbles slide down and accumulate at the bottom. The cost of screening by hand is the cost of shoveling the gravel against the screen divided by the number of cubic yards of saved material. In screening gravel for sand the richer the gravel is in fine material the cheaper will be the cost per cubic yard for screening; on the contrary in screening gravel for the pebbles the less sand there is in the gravel the cheaper will be the cost per cubic yard for screening. The cost of shoveling divided by the number of cubic yards shoveled is the cost of screening only when both the sand and the coarser material are saved. Tests made in the pit will enable the contractor to estimate how many cubic yards of gravel must be shoveled to get a cubic yard of sand or pebbles. An energetic man will shovel about 25 cu. yds. of gravel against a screen per 10-hour day and keep the screened material cleared away, providing no carrying is necessary

A mechanical arrangement capable of handling a considerably larger yardage of material is shown by Fig. 8. Two men and a team are required. The team is attached to the scraper by means of the rope passing through the pulley at the top of the incline. The scraper is loaded in the usual manner, hauled up the incline until its wheels are stopped by blocks and then the team is backed up to slacken the rope and permit the scraper to tip and dump its load. The trip holding the scraper while dumping is operated from the ground. The scraper load falls onto an inclined screen which takes out the sand and delivers the pebbles into the wagon. By erecting bins to catch the sand and pebbles this same arrangement could be made continuous in operation.

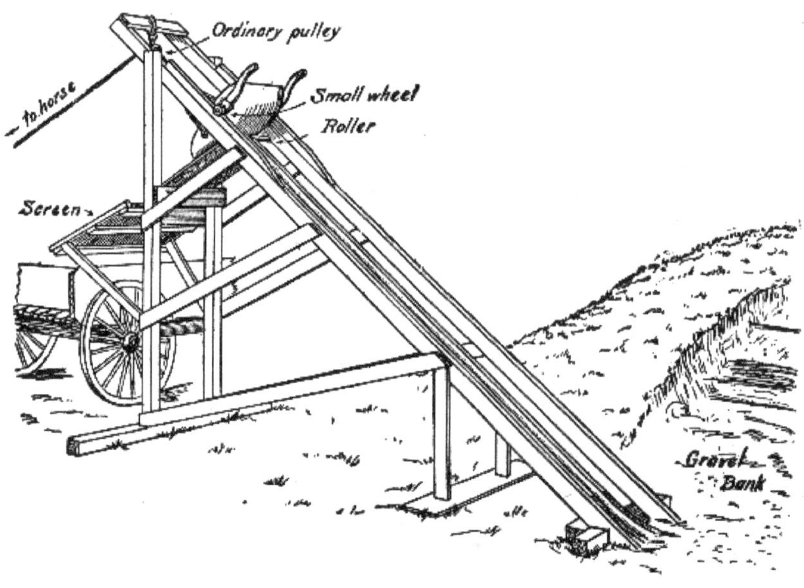

Fig. 8.—Device for Excavating and Screening Gravel and Loading Wagons.

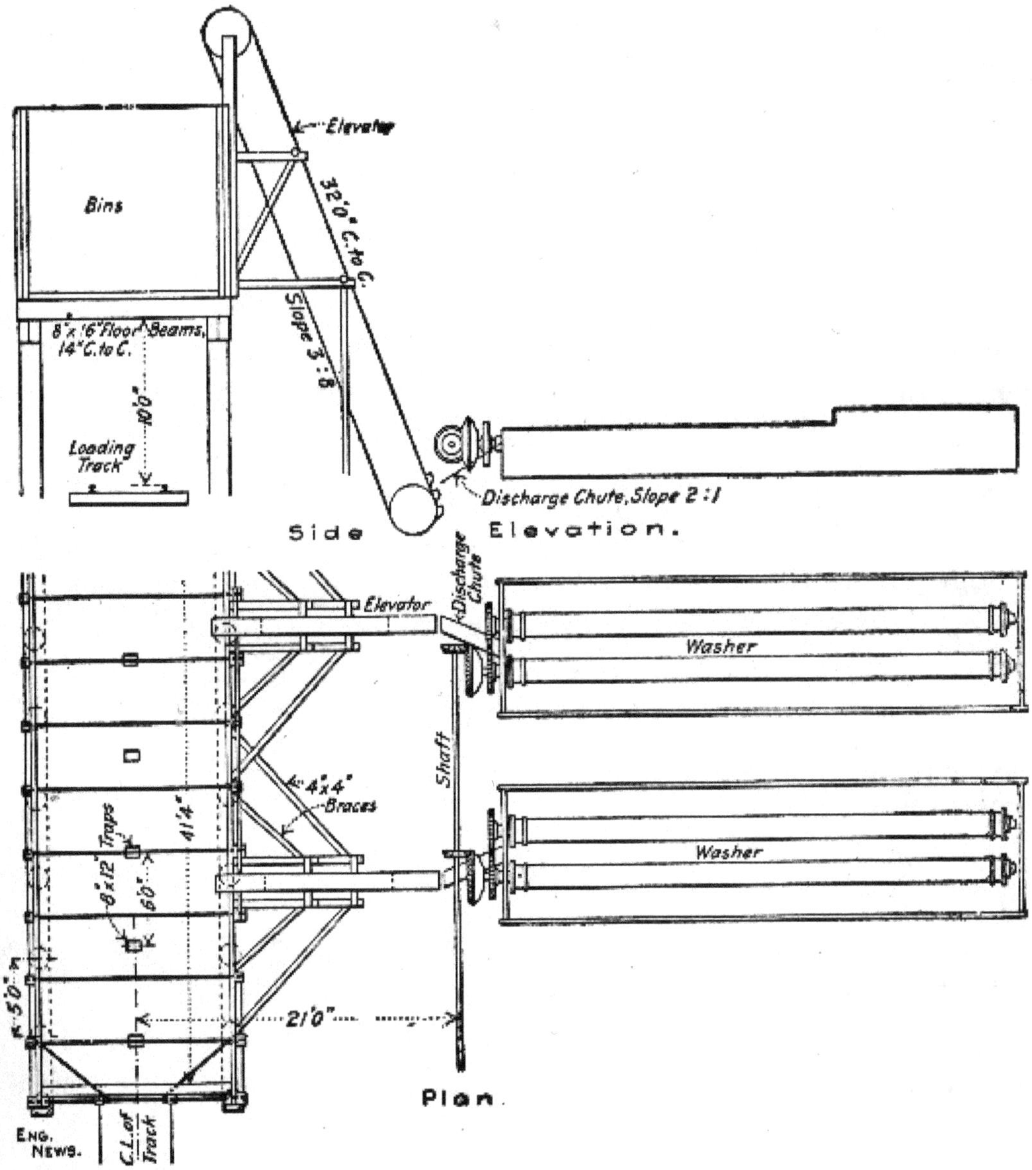

Fig. 9.—Gravel Washing Plant of 120 to 130 Cu. Yds., Per Hour Capacity.

In commercial gravel mining, the gravel is usually sorted into several sizes and generally it is washed as well as screened. Where the pebbles run into larger sizes a crushing plant is also usually installed to reduce the large stones. Works producing several hundred cubic yards of screened and washed gravel per day require a plant of larger size and greater cost than even a very large piece of concrete work will warrant, so that only general mention will be made here of such plants. The commercial sizes of gravel are usually 2-in., 1-in., ½-in. and ¼-in., down to sand. No very detailed costs of producing gravel by these commercial plants are available. At the plant of the Lake Shore & Michigan Southern Ry., where gravel is screened and washed for ballast, the gravel is passed over a 2-in., a ¾-in., a ¼-in. and a ⅛-in. screen in turn and the fine sand is saved. About 2,000 tons are handled per day; the washed gravel, 2-in. to ⅛-in. sizes, represents from 40 to 65 per cent. of the raw gravel and costs from 23 to 30 cts. per cu. yd., for excavation, screening and washing. The drawings of Fig. 9 show a gravel washing plant having a capacity of 120 to 130 cu. yds. per hour, operated by the Stewart-Peck Sand Co., of Kansas City, Mo. Where washing alone is necessary a plant of one or two washer units like those here shown could be installed without excessive cost by a contractor at any point where water is available. Each washer unit consists of two hexagonal troughs 18 ins. in diameter and 18 ft. long. A shaft carrying blades set spirally is rotated in each trough to agitate the gravel and force it along; each trough also has a fall of 6 ins. toward its receiving end. The two troughs are inclosed in a tank or box and above and between them is a 5-in. pipe having ¾-in. holes 3 ins. apart so arranged that the streams are directed into the troughs. The water and dirt pass off at the lower end of the troughs while the gravel is fed by the screws into a chute discharging into a bucket elevator, which in turn feeds into a storage bin. The gravel to be washed runs from 2 ins. to ⅛-in. in size; it is excavated by steam shovel and loaded into 1½ cu. yd. dump cars, three of which are hauled by a mule to the washers, where the load is dumped into the troughs. The plant having a capacity of 120 to 130 cu. yds. per hour cost $25,000, including pump and an 8-in. pipe line a mile long. A 100-hp. engine operates the plant, and 20 men are needed for all purposes. This plant produces washed gravel at a profit for 40 cts. per cu. yd.

2

CONCRETE PROPORTIONING

American engineers proportion concrete mixtures by measure, thus a 1-3-5 concrete is one composed of 1 volume of cement, 3 volumes of sand and 5 volumes of aggregate. In Continental Europe concrete is commonly proportioned by weight and there have been prominent advocates of this practice among American engineers. It is not evident how such a change in prevailing American practice would be of practical advantage. Aside from the fact that it is seldom convenient to weigh the ingredients of each batch, sand, stone and gravel are by no means constant in specific gravity, so that the greater exactness of proportioning by weight is not apparent. In this volume only incidental attention is given to gravimetric methods of proportioning concrete.

VOIDS.—Both the sand and the aggregates employed for concrete contain voids. The amount of this void space depends upon a number of conditions. As the task of proportioning concrete consists in so proportioning the several materials that all void spaces are filled with finer material the conditions influencing the proportion of voids in sand and aggregates must be known.

Voids in Sand.—The two conditions exerting the greatest influence on the proportion of voids in sand are the presence of moisture and the size of the grains of which the sand is composed.

TABLE I.—SHOWING EFFECT OF ADDITIONS OF DIFFERENT PERCENTAGES OF MOISTURE ON VOLUME OF SAND.

Per cent of water in sand	0	0.5	1	2	3	5	10
	Lbs.	Lbs.	Lbs.	Lbs.	Lbs.	Lbs.	Lbs.
Weight per cu. yd. of fine sand	3,457	2,206	2,085	2,044	2,037	2,035	2,133

and water	
Weight per cu. yd. of coarse sand and water	2,551 2,466 2,380 2,122 2,058 2,070 2,200

The volume of sand is greatly affected by the presence of varying percentages of moisture in the sand. A dry loose sand that has 45 per cent. voids if mixed with 5 per cent. by weight of water will swell, unless tamped, to such an extent that its voids may be 57 per cent. The same sand if saturated with water until it becomes a thin paste may show only 37½ per cent. voids after the sand has settled. Table I shows the results of tests made by Feret, the French experimenter. Two kinds of sand were used, a very fine sand and a coarse sand. They were measured in a box that held 2 cu. ft. and was 8 ins. deep, the sand being shoveled into the box but not tamped or shaken. After measuring and weighing the dry sand 0.5 per cent. by weight of water was added and the sand was mixed and shoveled back into the box again and then weighed. These operations were repeated with varying percentages of water up to 10 per cent. It will be noted that the weight of mixed water and sand is given; to ascertain the exact weight of dry sand in any mixture, divide the weight given in the table by 100 per cent. plus the given tabular per cent.; thus the weight of dry, fine sand in a 5 per cent. mixture is 2,035 ÷ 1.5 = 1,98 lbs. per cu. yd. The voids in the dry sand were 45 per cent. and in the sand with 5 per cent. moisture they were 56.7 per cent. Pouring water onto loose, dry sand compacts it. By mixing fine sand and water to a thin paste and allowing it to settle, it was found that the sand occupied 11 per cent. less space than when measured dry. The voids in fine sand, having a specific gravity of 2.65, were determined by measurement in a quart measure and found to be as follows:

Sand not packed, per cent. voids	44½
Sand shaken to refusal, per cent. voids	35
Sand saturated with water, per cent. voids	37½

Another series of tests made by Mr. H. P. Boardman, using Chicago sand having 34 to 40 per cent. voids, showed the following results:

Water added, per cent.	2	4	6	8	10
Resulting per cent. increase	17.6	22	19.5	16.6	15.6

Mr. Wm. B. Fuller found by tests that a dry sand, having 34 per cent. voids, shrunk 9.6 per cent. in volume upon thorough tamping until it had 27 per cent. voids. The same sand moistened with 6 per cent. water and loose had 44 per cent. voids, which was reduced to 31 per cent. by ramming. The same sand saturated with water had 33 per cent. voids and by thorough ramming its volume was reduced 8½ per cent. until the sand had only 26¼ per cent. voids. Further experiments might be quoted and will be found recorded in several general treatises on concrete, but these are enough to demonstrate conclusively that any theory of the quantity of cement in mortar to be correct must take into account the effect of moisture on the voids in sand.

The effect of the size and the shape of the component grains on the amount of voids in sand is considerable. Feret's experiments are conclusive on these points, and they alone will be followed here. Taking for convenience three sizes of sand Feret mixed them in all the varying proportions possible with a total of 10 parts; there were 66 mixtures. The sizes used were: Large (*L*), sand composed of grains passing a sieve of 5 meshes per linear inch and retained on a sieve of 15 meshes per linear inch; medium (*M*), sand passing a sieve of 15 meshes and retained on a sieve of 50 meshes per linear inch, and fine (*F*), sand passing a 50-mesh sieve. With a dry sand whose grains have a specific gravity of 2.65, the weight of a cubic yard of either the fine, or the medium, or the large size, was 2,190 lbs., which is equivalent to 51 per cent. voids. The greatest weight of mixture, 2,840 lbs. per cu. yd., was an $L_6M_0F_4$ mixture, that is, one composed of six parts large, no parts medium and 4 parts fine; this mixture was the densest of the 66 mixtures made, having 36 per cent. voids. It will be noted that the common opinion that the densest mixture is obtained by a mixture of gradually increasing sizes of grains is incorrect; there must be enough difference in the size of the grains to provide voids so large that the smaller grains will enter them and not wedge the larger grains apart. Turning now to the shape of the grains, the tests showed that rounded grains give less voids than angular grains. Using sand having a composition of $L_5M_3F_2$ Feret got the following results:

	—Per cent. Voids—	
Kind of Grains.	Shaken.	Unshaken.

Natural sand, rounded grains	25.6	35.9
Crushed quartzite, angular grains	27.4	42.1
Crushed shells, flat grains	31.8	44.3
Residue of quartzite, flat grains	34.6	47.5

The sand was shaken until no further settlement occurred. It is plain from these data on the effect of size and shape of grains on voids why it is that discrepancies exist in the published data on voids in dry sand. An idea of the wide variation in the granulometric composition of different sands is given by Table II. Table III shows the voids as determined for sands from different localities in the United States.

TABLE II.—SHOWING GRANULOMETRIC COMPOSITIONS OF DIFFERENT SANDS.

Held by a Sieve.	A	B	C	E
No. 10	35.3%			
No. 20	32.1	12.8%	4.2%	11%
No. 30	14.6	49.0	12.5	14
No. 40	...	...	44.4	...
No. 50	9.6	29.3	...	53
No. 100	4.9	5.7	...	...
No. 200	2.0	2.3	...	...
Voids	33%	39%	41.7%	31%

NOTE.—A, is a "fine gravel" (containing 8% clay) used at Philadelphia. B, Delaware River sand. C, St. Mary's River sand. D, Green River, Ky., sand, "clean and sharp."

TABLE III.—SHOWING MEASURED VOIDS IN SAND FROM DIFFERENT LOCALITIES.

Locality.	Authority.	Percent Voids.	Remarks.

Ohio River	W. M. Hall	31	Washed
Sandusky, O.	C. E. Sherman	40	Lake
Franklin Co., O.	C. E. Sherman	40	Bank
Sandusky Bay, O.	S. B. Newberry	32.3	......
St. Louis, Mo.	H. H. Henby	34.3	Miss. River
Sault Ste. Marie	H. von Schon	41.7	River
Chicago, Ill.	H. P. Broadman	34 to 40	......
Philadelphia, Pa	39	Del. River	
Mass. Coast	31 to 34	......	
Boston, Mass	Geo. Kimball	33	Clean
Cow Bay, L. I.	Myron S. Falk	40½	......
Little Falls, N. J.	W. B. Fuller	45.6	......
Canton, Ill.	G. W. Chandler	30	Clean

Voids in Broken Stone and Gravel.—The percentage of voids in broken stone varies with the nature of the stone: whether it is broken by hand or by crushers; with the kind of crusher used, and upon whether it is screened or crusher-run product. The voids in broken stone seldom exceed 52 per cent. even when the fragments are of uniform size and the stone is shoveled loose into the measuring box. The following records of actual determinations of voids in broken stone cover a sufficiently wide range of conditions to show about the limits of variation.

The following are results of tests made by Mr. A. N. Johnson, State Engineer of Illinois, to determine the variation in voids in crushed stone due to variation in size and to method of loading into the measuring box. The percentage of voids was determined by weighing the amount of water added to fill the box:

Size.	Method of Loading.	Per cent. of Voids.
3 in.	20-ft. drop	41.8
3 in.	15-ft drop	46.8
3 in.	15-ft. drop	47.2
3 in.	Shovels	48.7
1½ in.	20-ft. drop	42.5

1½ in.	15-ft. drop	46.8
1½ in.	15-ft. drop	46.8
1½ in.	Shovels	50.5
¾ in.	20-ft. drop	39.4
¾ in.	15-ft. drop	42.7
¾ in.	15-ft. drop	41.5
¾ in.	15-ft. drop	41.8
¾ in.	Shovels	45.2
¾ in.	Shovels	44.6
⅜ in.	Shovels	41.0
⅜ in.	Shovels	40.6
⅜ in.	Shovels	41.0

The table shows clearly the effect on voids of compacting the stone by dropping it; it also shows for the ¾-in. and the ⅜-in. stone loaded by shovels how uniformly the percentages of voids run for stone of one size only. Dropping the stone 20 ft. reduced the voids some 12 to 15 per cent. as compared with shoveling.

TABLE IV.—SHOWING DETERMINED PERCENTAGES OF VOIDS IN BROKEN STONE FROM VARIOUS COMMON ROCKS.

Authority.	Percent Voids.	Remarks.
Sabin	49.0	Limestone, crusher run after screening out ⅛-in. and under.
"	44.0	Limsetone (1 part screenings mixed with 6 parts broken stone).
Wm. M. Black	46.5	Screened and washed, 2-ins. and under.
J. J. R. Croes	47.5	Gneiss, after screening out ¼-in. and under.
S. B. Newberry	47.0	Chiefly about egg size.
H. P. Broadman	39 to 42	Chicago limestone, crusher run.
"	48 to 52	" " screened into sizes.

Wm. M. Hall	48.0	Green River limestone, 2½-ins. and smaller dust screened out.
"	50.0	Hudson River trap, 2½-ins. and smaller, dust screened out.
Wm. B. Fuller	47.6	New Jersey trap, crusher run, 1/6 to 2.1 in.
Geo. A. Kimball	49.5	Roxbury conglomerate, ½ to 2½ ins.
Myron S. Falk	48.0	Limestone, ½ to 3 ins.
W. H. Henby	43.0	" 2-in size.
"	46.0	" 1½-in size
Feret	53.4	Stone, 1.6 to 2.4 ins.
"	51.7	" 0.8 to 1.6 in.
"	52.1	" 0.4 to 0.8 in.
A. W. Dow	45.3	Bluestone, 89% being 1½ to 2½ ins.
"	45.3	" 90% being 1/6 to 1½ in.
Taylor and Thompson	54.5	Trap, hard, 1 to 2½ ins.
"	54.5	" " ½ to 1 in.
"	45.0	" " 0 to 2½ in.
"	51.2	" soft, ¾ to 2 ins.
G. W. Chandler	40.0	Canton, Ill.
Emile Low	39.0	Buffalo limestone, crusher run, dust in.
C. M. Saville	46.0	Crushed cobblestone, screened into sizes.

Table V.—Showing Percentages of Voids in Gravel and Broken Stone of Different Granulometric Compositions.

				Per cent Voids in	
Passing a ring of	2.4"	1.6"	0.8"	Round	Broken
Held by a ring	1.6"	0.8"	0.4"	Pebbles.	Stone.
Parts	1	0	0	40.0	53.4

"	0	1	0	38.8	51.7
"	0	0	1	41.7	52.1
"	1	1	0	35.8	50.5
"	1	0	1	35.6	47.1
"	0	1	1	37.9	40.5
"	1	1	1	35.5	47.8
"	4	1	1	34.5	49.2
"	1	4	1	36.6	49.4
"	1	1	4	38.1	48.6
"	8	0	2	34.1	....

Table IV gives the voids in broken stone as determined by various engineers; it requires no explanation. Table V, taken from Feret's tests, shows the effect of changes in granulometric composition on the amount of voids in both broken stone and gravel. Considering the column giving voids in stone it is to be noted first how nearly equal the voids are for stone of uniform size whatever that size be. As was the case with sand a mixture of coarse and fine particles gives the fewest voids; for stone an $L_1M_0F_1$ mixture and for gravel an $L_8M_0F_2$ mixture. Tamping reduces the voids in broken stone. Mr. Geo. W. Rafter gives the voids in clean, hand-broken limestone passing a 2½-in. ring as 43 per cent. after being lightly shaken and 37½ per cent. after being rammed. Generally speaking heavy ramming will reduce the voids in loose stone about 20 per cent.

It is rare that gravel has less than 30 per cent. or more than 45 per cent. voids. If the pebbles vary considerably in size so that the small fit in between the large, the voids may be as low as 30 per cent. but if the pebbles are tolerably uniform in size the voids will approach 45 per cent. Table V shows the effect of granulometric composition on the voids in gravel as determined by Feret. Mr. H. Von Schon gives the following granulometric analysis of a gravel having 34.1 per cent. voids:

Retained on 1-in. ring, per cent.	10.70
Retained on ⅜-in. ring, per cent.	23.65
Retained on No. 4 sieve, per cent.	8.70

Retained on No. 10 sieve, per cent.	17.14
Retained on No. 20 sieve, per cent.	21.76
Retained on No. 30 sieve, per cent.	6.49
Retained on No. 40 sieve, per cent.	5.96
Passed a No. 40 sieve, per cent.	5.59
Passed a 1½-in ring, per cent.	100.00

As mixtures of broken stone and gravel are often used the following determinations of voids in such mixtures are given. The following determinations were made by Mr. Wm. M. Hall for mixtures of blue limestone and Ohio River washed gravel:

Per cent. Stone.		Per cent. Gravel.	Per cent. Voids in Mix
100	with	0	48
80	"	20	44
70	"	30	41
60	"	40	38½
50	"	50	36
0	"	100	35

The dust was screened from the stone all of which passed a 2½-in. ring; the gravel all passed a 1½-in. screen. Using the same sizes of gravel and Hudson River trap rock, the results were:

Per cent. Trap.		Per cent. Gravel.	Per cent. Voids in Mix.
100	with	0	50
60	"	40	38½
50	"	50	36
0	"	100	35

The weight of a cubic foot of loose gravel or stone is not an accurate index of the percentage of voids unless the specific gravity is known. Pure quartz weighs 165 lbs., per cu. ft., hence broken quartz having 40 per cent. voids weighs 165 × .60 = 99 lbs. per cu. ft. Few gravels are entirely quartz, and many contain stone having a greater specific gravity like some traps or a less specific gravity like some shales and sandstone. Tables VI and VII give the specific gravities of common stones and minerals and Table VIII gives the weights corresponding to different percentages of voids for different specific gravities.

TABLE VI.—SPECIFIC GRAVITY OF STONE. (Condensed from Merrill's "Stones for Building.")

Trap,	Boston, Mass.	2.78
"	Duluth, Minn.	2.8 to 3.0
"	Jersey City, N. J.	3.03
"	Staten Island, N. Y.	2.86
Gneiss,	Madison Ave., N. Y.	2.92
Granite,	New London, Conn.	2.66
"	Greenwich, Conn.	2.84
"	Vinalhaven, Me.	2.66
"	Quincy, Mass.	2.66
"	Barre, Vt.	2.65
Limestone,	Joliet, Ill.	2.56
"	Quincy, Ill.	2.51 to 2.57
Limestone, (oolitic)	Bedford, Ind.	2.25 to 2.45
"	Marquette, Mich.	2.34
"	Glens Falls, N.Y.	2.70
"	Lake Champlain, N. Y.	2.75
Sandstone,	Portland, Conn.	2.64
"	Haverstraw, N. Y.	2.13
"	Medina, N. Y.	2.41
"	Potsdam, N. Y.	2.60
"	(grit) Berea, O.	2.12

TABLE VII.—SPECIFIC GRAVITY OF COMMON MINERALS AND ROCKS.

Apatite	2.92-3.25
Basalt	3.01
Calcite, $CaCO_3$	2.5-2.73

Cassiterite, SnO_2	6.4-7.1
Cerrusite, $PbCO_3$	6.46-6.48
Chalcopyrite, $CuFeS_2$	4.1-4.3
Coal, anthracite	1.3-1.84
Coal, bituminous	1.2-1.5
Diabase	2.6-3.03
Diorite	2.92
Dolomite, CaMg $(CO_3)^2$	2.8-2.9
Felspar	2.44-2.78
Felsite	2.65
Galena, Pbs	7.25-7.77
Garnet	3.15-4.31
Gneiss	2.62-2.92
Granite	2.55-2.86
Gypsum	2.3-3.28
Halite (salt) NaCl	2.1-2.56
Hematite, Fe_2O_3	4.5-5.3
Hornblende	3.05-3.47
Limonite, $Fe_3O_4\ (OH)^6$	3.6-4.0
Limestone	2.35-2.87
Magnetite, Fe_3O_4	4.9-5.2
Marble	2.08-2.85
Mica	2.75-3.1
Mica Schist	2.5-2.9
Olivine	3.33-3.5
Porphyry	2.5-2.6
Pyrite, FeS_2	4.83-5.2
Quartz, SiO_2	2.5-2.8
Quartzite	2.6-2.7
Sandstone	2.0-2.78
" Medina	2.4
" Ohio	2.2
" Slaty	1.82

Shale	2.4-2.8
Slate	2.5-2.8
Sphalerite, ZnS	3.9-4.2
Stibnite, Sb_2S_3	4.5-4.6
Syenite	2.27-2.65
Talc	2.56-2.8
Trap	2.6-3.0

Table VIII.—Showing Weight of Stone with Different Percentages of Voids for Different Specific Gravities.

Specific Gravity.	Weight in Lbs. per cu. ft.	Weight in Lbs. per cu. yd.	Weight in Lbs. per cu. yd. when Voids are				
			30%	35%	40%	45%	50%
1.0	62.355	1,684	1,178	1,094	1,010	926	842
2.0	124.7	3,367	2,357	2,187	2,020	1,852	1,684
2.1	130.9	3,536	2,475	2,298	2,121	1,945	1,768
2.2	137.2	3,704	2,593	2,408	2,222	2,037	1,852
2.3	143.4	3,872	2,711	2,517	2,323	2,130	1,936
2.4	149.7	4,041	2,828	2,626	2,424	2,222	2,020
2.5	155.9	4,209	2,946	2,736	2,525	2,315	2,105
2.6	162.1	4,377	3,064	2,845	2,626	2,408	2,189
2.7	168.4	4,546	3,182	2,955	2,727	2,500	2,273
2.8	174.6	4,714	3,300	3,064	2,828	2,593	2,357
2.9	180.9	4,882	3,418	3,174	2,929	2,685	2,441
3.0	187.1	5,051	3,536	3,283	3,030	2,778	2,526
3.1	193.3	5,219	3,653	3,392	3,131	2,871	2,609
3.2	199.5	5,388	3,771	3,502	3,232	2,963	2,694
3.3	205.8	5,556	3,889	3,611	3,333	3,056	2,778
3.4	212.0	5,724	4,007	3,721	3,434	3,148	2,862
3.5	218.3	5,893	4,125	3,830	3,535	3,241	2,947

In buying broken stone by the cubic yard it should be remembered that hauling in a wagon compacts the stone by shaking it down and reduces the volume. Table IX shows the results of tests made by the Illinois Highway Commission to determine the settlement of crushed stone in wagon loads for different lengths of haul. The road over which the tests were made was a macadam road, not particularly smooth, but might be considered as an average road surface. The wagon used was one with a dump bottom supported by chains, which were drawn as tight as possible, so as to reduce the sag to a minimum. It will be noticed that about 50 per cent. of the settlement occurs within the first 100 ft., and 75 per cent. of the settlement in the first 200 ft. Almost all of the settlement occurs during the first half mile, as the tests showed practically no additional settlement for distances beyond. Some of the wagons were loaded from the ground with shovels, others were loaded from bins, the stone having a 15-ft. drop, which compacted the stone a little more than where loaded with shovels, so that there was somewhat less settlement. But at the end of a half mile the density was practically the same, whatever the method of loading. The density at the beginning and at the end of the haul can be compared by the weight of a given volume of crushed stone. For convenience, the weight of a cubic yard of the material at the beginning of the haul and at the end was computed from the known contents of a wagon.

TABLE IX.—SHOWING SETTLEMENT OF BROKEN STONE DUE TO DIFFERENT LENGTHS OF HAUL ON ORDINARILY GOOD ROAD IN WAGONS.

		Per cent Settlement for Hauling.									Weight per Cu. Yd. in Lbs.	
Size.	Method of Loading.	100'	200'	300'	400'	500'	600'	700'	½ Mile	1 Mile	At start.	At finish.
Screenings	15 ft. drop	....	...	....	....	....	....	....	11.5	11.5	2,518	2,840
Screenings	15 ft. drop	....	...	....	....	....	....	....	12.6	12.6	2,518	2,886

Screenings	15 ft. drop	7.3	8.3	8.9	9.2	9.5	10.1	10.1	11.2	....	2,450	2,770
Screenings	15 ft. drop	5.0	9.6	10.2	10.2	10.4	10.4	10.4	12.4	....	,425	2,780
1½ inch	15 ft. drop	...	....	....	....	....	....	....	11.5	11.5[C]	2,305	2,600
1½ inch	15 ft. drop	5.3	6.2	7.1	7.7	7.9	8.0	8.3	9.2	....	2,380	2,625
1½ inch	15 ft. drop	2.6	3.7	4.9	5.3	5.3	5.3	5.4	5.4	....	2,450	2,600
1½ inch	Shovels	3.5	4.1	4.8	5.3	5.3	5.7	6.5	7.25	....	2,270	2,445
1½ inch	Shovels	...	....	....	....	....	....	....	12.6	12.6	2,305	2,642
3 inch	15 ft. drop	...	....	....	....	....	....	....	10.1	10.1	2,376	2,638
3 inch	15 ft. drop	3.5	4.2	4.5	4.8	5.0	5.0	5.0	6.0	....	2,360	2,505
3 inch	15 ft. drop	0.5	2.5	2.5	4.1	4.3	4.3	4.3	4.9	....	2,470	2,595
3 inch	Shovels	...	....	....	....	....	....	....	12.6	12.6	2,270	2,601
3 inch	Shovels	5.0	5.6	6.5	6.5	6.8	6.8	6.8	7.1	....	2,335	2,510

[C] —Same per cent of settlement for two-mile haul.

THEORY OF THE QUANTITY OF CEMENT IN MORTAR AND CONCRETE.—All sand contains a large percentage of voids; in 1 cu. ft. of loose sand there is 0.3 to 0.5 cu. ft. of voids, that is, 30 to 50 per cent. of the sand is voids. In making mortar the cement is mixed with the sand and the flour-like particles of the cement fit in between the grains of sand occupying a part or all of the voids. The amount of cement required in a mortar will naturally depend upon the amount of voids in the particular sand with which it is mixed and since a correct estimate of the number of barrels of cement per cubic yard of mortar is very important, and since it is not always possible to make actual mixtures before bidding, rules based on various theories have been formulated for determining these quantities. In this volume the rule based on the theory outlined by one of the authors in 1901 will be followed. The following is a discussion of the authors' theory:

When loose sand is mixed with water, its volume or bulk is increased; subsequent jarring will decrease its volume, but still leave a net gain of about 10 per cent.; that is, 1 cu. ft. of dry sand becomes about 1.1 cu. ft. of damp sand. Not only does this increase in the volume of the sand occur, but, instead of increasing the voids that can be filled with cement, there is an absolute loss in the volume of available voids. This is due to the space occupied by the water necessary to bring the sand to the consistency of mortar; furthermore, there is seldom a perfect mixture of the sand and cement in practice, thus reducing the available voids. It is safe to call this reduction in available voids about 10 per cent.

When loose, dry Portland cement is wetted, it shrinks about 15 per cent, in volume, behaving differently from the sand, but it never shrinks back to quite as small a volume as it occupies when packed tightly in a barrel. Since barrels of different brands vary widely in size, the careful engineer or contractor will test any brand he intends using in large quantities, in order to ascertain exactly how much cement paste can be made. He will find a range of from 3.2 cu. ft. to 3.8 cu. ft. per barrel of Portland cement. Obviously the larger barrel may be cheaper though its price is higher. Specifications often state the number of cubic feet that will be allowed per barrel in mixing the concrete ingredients, so that any rule or formula to be of practical value must contain a factor to allow for the specified size of the barrel, and another factor to allow for the actual number of cubic feet of paste that a barrel will yield—the two being usually quite different.

The deduction of a rational, practical formula for computing the quantity of cement required for a given mixture will now be given, based upon the facts above outlined.

Let p = number of cu. ft. cement paste per bbl., as determined by actual test.
n = number of cu. ft. of cement per bbl., as specified in the specifications.
s = parts of sand (by volume) to one part of cement, as specified.
g = parts of gravel or broken stone (by volume) to one part of cement, as specified.
v = percentage of voids in the dry sand, as determined by test.
V = percentage of voids in the gravel or stone, as determined by test.

Then, in a mortar of 1 part cement to s parts sand, we have:

$n\,s$ = cu. ft. of dry sand to 1 bbl. of cement.
$n\,s\,v$ = " " " voids in the dry sand.
$0.9\,n\,s\,v$ = " " " available voids in the wet sand.

$$1.1\,n\,s = \text{" " " wet sand.}$$
$$p - 0.9\,n\,s\,v = \text{" " " cement paste in excess of the voids.}$$

Therefore:

$$1.1\,n\,s + (p - 0.9\,n\,s\,v) = \text{cu. ft. of mortar per bbl.}$$

Therefore:

$$N = \frac{27}{1.1\,n\,s + (p - 0.9\,n\,s\,v)} = \frac{27}{p + n\,s\,(1.1 - 0.9\,v)}$$

N being the number of barrels of cement per cu. yd. of mortar.

When the mortar is made so lean that there is not enough cement paste to fill the voids in the sand, the formula becomes:

$$N = \frac{27}{1.1\,n\,s}$$

A similar line of reasoning will give us a rational formula for determining the quantity of cement in concrete; but there is one point of difference between sand and gravel (or broken stone), namely, that the gravel does not swell materially in volume when mixed with water. However, a certain amount of water is required to wet the surface of the pebbles, and this water reduces the available voids, that is, the voids that can be filled by the mortar. With this in mind, the following deduction is clear, using the nomenclature and symbols above given:

$ng =$	cu.	ft.	of	dry gravel (or stone).
$ng\,V =$	"	"	"	voids in dry gravel.
$0.9\,ng\,V =$	"	"	"	"available voids" in the wet gravel.
$p + n\,s\,(1.1 - 0.9\,v) - 0.9\,ng\,V =$ excess of mortar				over the available voids in the wet gravel.
$ng + p + n\,s\,(1.1 - 0.9\,v) - 0.9\,ng\,V =$	cu.	ft.	of	concrete from 1 bbl. cement.

$$N = \frac{27}{p + n\,s\,(1.1 - 0.9v) + ng\,(1 - 0.9\,V)}$$

N being the number of barrels of cement required to make 1 cu. yd. of concrete.

This formula is rational and perfectly general. Other experimenters may find it desirable to use constants slightly different from the 1.1 and the 0.9, for fine sands swell more than coarse sands, and hold more water.

The reader must bear in mind that when the voids in the sand exceed the cement paste, and when the available voids in the gravel (or stone) exceed the mortar, the formula becomes:

$$N = \frac{27}{ng}$$

These formulas give the amounts of cement in mortars and concretes compacted in place. Tables X to XIII are based upon the foregoing theory, and will be found to check satisfactorily with actual tests.

In using these tables remember that the proportion of cement to sand is by volume, and not by weight. If the specifications state that a barrel of cement shall be considered to hold 4 cu. ft., for example, and that the mortar shall be 1 part cement to 2 parts sand, then 2 barrel of cement is mixed with 8 cu. ft. of sand, regardless of what is the actual size of the barrel, and regardless of how much cement paste can be made with a barrel of cement. If the specifications fail to state what the size of a barrel will be, then the contractor is left to guess.

TABLE X.—BARRELS OF PORTLAND CEMENT PER CUBIC YARD OF MORTAR.

(Voids in sand being 35%, and 1 bbl. cement yielding 3.65 cu. ft. of cement paste.)

Proportion of Cement to Sand	1 to 1	1 to 1½	1 to 2	1 to 2½	1 to 3	1 to 4
	Bbls.	Bbls.	Bbls.	Bbls.	Bbls.	Bbls.
Barrel specified to be 3.5 cu. ft.	4.22	3.49	2.97	2.57	2.28	1.76
" " " 3.8 "	4.09	3.33	2.81	2.45	2.16	1.62
" " " 4.0 "	4.00	3.24	2.73	2.36	2.08	1.54
" " " 4.4 "	3.81	3.07	2.57	2.27	2.00	1.40
Cu. yds. sand per cu. yd. mortar	0.6	0.7	0.8	0.9	1.0	1.0

TABLE XI.—BARRELS OF PORTLAND CEMENT PER CUBIC YARD OF MORTAR.

(Voids in sand being 45%, and 1 bbl. cement yielding 3.4 cu. ft. of cement paste.)

Proportion of Cement to Sand	1 to 1	1 to 1½	1 to 2	1 to 2½	1 to 3	1 to 4
	Bbls.	Bbls.	Bbls.	Bbls.	Bbls.	Bbls.
Barrel specified to be 3.5 cu. ft.	4.62	3.80	3.25	2.84	2.35	1.76
" " " 3.8 "	4.32	3.61	3.10	2.72	2.16	1.62
" " " 4.0 "	4.19	3.46	3.00	2.64	2.05	1.54
" " " 4.4 "	3.94	3.34	2.90	2.57	1.86	1.40
Cu. yds. sand per cu. yd. mortar	0.6	0.8	0.9	1.0	1.0	1.0

If the specifications call for proportions by weight, assume a Portland barrel to contain 380 lbs. of cement, and test the actual weight of a cubic foot of the sand to be used. Sand varies extremely in weight, due both to the variation in the per cent. of voids, and to the variation in the kind of minerals of which the sand is composed. A quartz sand having 35 per cent. voids weighs 107 lbs. per cu. ft.; but a quartz sand having 45 per cent. voids weighs only 91 lbs. per cu. ft. If the weight of the sand must be guessed at, assume 100 lbs. per cu. ft. If the specifications require a mixture of 1 cement to 2 of sand by weight, we will have 380 lbs. (or 1 bbl.) of cement mixed with 2 × 380, or 760 lbs. of sand; and if the sand weighs 90 lbs. per cu. ft., we shall have 760 ÷ 90, or 8.44 cu. ft. of sand to every barrel of cement. In order to use the tables above given, we may specify our own size of barrel; let us say 4 cu. ft.; then 8.44 ÷ 4 gives 2.11 parts of sand by volume to 1 part of cement. Without material error we may call this a 1 to 2 mortar, and use the tables, remembering that our barrel is now "specified to be" 4 cu. ft. If we have a brand of cement that yields 3.4 cu. ft. of paste per bbl., and sand having 45 per cent. voids, we find that approximately 3 bbls. of cement per cu. yd. of mortar will be required.

Table XII.—Ingredients in 1 Cubic Yard of Concrete.

(Sand voids, 40%; stone voids, 45%; Portland cement barrel yielding 3.65 cu. ft. paste. Barrel specified to be 3.8 cu. ft.)

	1:2:4	1:2:5	1:2:6	1:2½:5	1:2½:6	1:3:4
Bbls. cement per cu. yd. concr't	1.46	1.30	1.18	1.13	1.00	1.25
Cu. yds. sand " "	0.41	0.36	0.33	0.40	0.35	0.53
Cu. yds. stone " "	0.82	0.90	1.00	0.80	0.84	0.71
Proportions by Volume.	1:3:5	1:3:6	1:3:7	1:4:7	1:4:8	1:4:9

Bbls. cement per cu. yd. concr't	1.13	1.05	0.96	0.82	0.77	0.73
Cu. yds. sand " "	0.48	0.44	0.40	0.46	0.43	0.41
Cu. yds. stone " "	0.80	0.88	0.93	0.80	0.86	0.92

NOTE.—This table is to be used where cement is measured packed in the barrel, for the ordinary barrel holds 3.8 cu. ft.

It should be evident from the foregoing discussions that no table can be made, and no rule can be formulated that will yield accurate results unless the brand of cement is tested and the percentage of voids in the sand determined. This being so the sensible plan is to use the tables merely as a rough guide, and, where the quantity of cement to be used is very large, to make a few batches of mortar using the available brands of cement and sand in the proportions specified. Ten dollars spent in this way may save a thousand, even on a comparatively small job, by showing what cement and sand to select.

It will be seen that Tables XII and XIII can be condensed into the following rule:

Add together the number of parts and divide this sum into ten, the quotient will be approximately the number of barrels of cement per cubic yard.

TABLE XIII.—INGREDIENTS IN 1 CUBIC YARD OF CONCRETE.

(Sand voids, 40%; stone voids, 45%; Portland cement barrel yielding 3.65 cu. ft. of paste. Barrel specified to be 4.4 cu. ft.)

Proportions by Volume.	1:2:4	1:2:5	1:2:6	1:2½:5	1:2½:6	1:3:4
Bbls. cement per cu. yd. concr't	1.30	1.16	1.00	1.07	0.96	1.08
Cu. yds. sand " "	0.42	0.38	0.33	0.44	0.40	0.53
Cu. yds. stone " "	0.84	0.95	1.00	0.88	0.95	0.71
Proportions by Volume.	1:3:5	1:3:6	1:3:7	1:4:7	1:4:8	1:4:9
Bbls. cement per cu. yd. concr't	0.96	0.90	0.82	0.75	0.68	0.64
Cu. yds. sand " "	0.47	0.44	0.40	0.49	0.44	0.42
Cu. yds. stone " "	0.78	0.88	0.93	0.86	0.88	0.95

NOTE.—This table is to be used when the cement is measured loose, after dumping it into a box, for under such conditions a barrel of cement yields 4.4 cu. ft. of loose cement.

Thus for a 1:2:5 concrete, the sum of the parts is 1 + 2 + 5, which is 8; then 10 ÷ 8 is 1.25 bbls., which is approximately equal to the 1.30 bbls. given in the table. Neither is this rule nor are the tables applicable if a different size of cement barrel is specified, or if the voids in the sand or stone differ materially from 40 per cent. to 45 per cent. respectively. There are such innumerable combinations of varying voids, and varying sizes of barrel, that the authors do not deem it worth while to give other tables. The following amounts of cement per cubic yard of mortar were determined by test:

Authority	Neat.	1 to 1	1 to 2	1 to 3	1 to 4	1 to 5	1 to 6	1 to 7	1 to 8
	Bbls.	Bbls.	Bbls.	Bbls.	Bbls.	Bbls.	Bbls.	Bbls.	Bbls.
Sabin	7.40	4.17	2.84	2.06	1.62	1.33	1.14	....	....
W. B. Fuller	8.02	4.58	3.09	2.30	1.80	1.48	1.23	1.11	1.00
H. P. Boardman.	7.40	4.50	3.18	2.35	....	....	....	....	....

The proportions were by barrels of cement to barrels of sand, and Sabin called a 380-lb. barrel 3.65 cu. ft., whereas Fuller called a 380-lb. barrel 3.80 cu. ft.; and Boardman called a 380-lb. barrel 3.5 cu. ft. Sabin used a sand having 38 per cent. voids; Fuller used a sand having 45 per cent. voids; and Boardman used a sand having 38 per cent. voids. It will be seen that the cement used by Sabin yielded 3.65 cu. ft. of cement paste per bbl. (i. e. 27 ÷ 7.4), whereas the (Atlas) cement used by Fuller yielded 3.4 cu. ft. of cement paste per bbl. Sabin found that a barrel of cement measured 4.37 cu. ft. when dumped and measured loose. Mr. Boardman states a barrel (380 lbs., net) of Lehigh Portland cement yields 3.65 cu. ft. of cement paste; and that a barrel (265 lbs., net) of Louisville natural cement yields 3.0 cu. ft. of cement paste.

Mr. J. J. R. Croes, M. Am. Soc. C. E., states that 1 bbl. of Rosendale cement and 2 bbl. of sand (8 cu. ft.) make 9.7 cu. ft. of mortar, the extreme variations from this average being 7 per cent.

Frequently concrete is made by mixing one volume of cement with a given number of volumes of pit gravel; no sand being used other than the sand that is found naturally mixed with the gravel. In such cases the cement rarely increases the bulk of the gravel, hence Table XIV will give the approximate amount of cement, assuming 1 cu. yd. of gravel per cubic yard of concrete.

Table XIV.—Showing Barrels of Cement per Cubic Yard of Various Mixtures of Cement and Pit Gravel.

Spc. Vol. of bbl. cu. ft.	Barrels of Cement per Cubic Yard of Concrete for Mixtures of						
	1-5	1-6	1-7	1-8	1-9	1-10	1-12
3.8	1.41	1.18	1.01	0.874	0.789	0.71	0.59
4.4	1.25	1.02	0.875	0.766	0.681	0.61	0.51

PERCENTAGE OF WATER IN CONCRETE.—Tests show that dry mixtures when carefully deposited and well tamped produce the stronger concrete. This superiority of dry mixtures it must be observed presupposes careful deposition and thorough tamping, and these are tasks which are difficult to have accomplished properly in actual construction work and which, if accomplished properly, require time and labor. Wet mixtures readily flow into the corners and angles of the forms and between and around the reinforcing bars with only a small amount of puddling and slicing and are, therefore, nearly always used because of the time and labor saved in depositing and tamping. The following rule by which to determine the percentage of water by weight for any given mixture of mortar for wet concrete will be found satisfactory:

Multiply the parts of sand by 8, add 24 to the product, and divide the total by the sum of the parts of sand and cement.

For example if the percentage of water is required for a 1-3 mortar:

$$\frac{(3 \times 8) + 24}{4} = 12.$$

Hence the water should be 12 per cent. of the combined weight of cement and sand. For a 1-1 mortar the rule gives 16 per cent.; for a 1-2 mortar it gives 13½ per cent., and for a 1-6 mortar it gives 10.3 per cent.

To calculate the amount of water per cubic yard of 1-3-6 concrete for example the procedure would be as follows: By the above rule a 1-3 mortar requires

$$\frac{(3 \times 8) + 24}{4} = 12 \text{ per cent. water.}$$

A 1-3-6 concrete, according to Table XII, contains 1.05 bbls. cement and 0.44 cu. yd. sand. Cement weighs 380 lbs. per barrel, hence 1.05 bbls. would weigh 380 × 1.05 = 399 lbs. Sand weighs 2,700 lbs. per cu. yd., hence 0.44 cu. yd. of sand would weigh 2,700 × 0.44 = 1,188 lbs. The combined weight of the cement and sand would thus be 399 + 1,188 = 1,587 lbs. and 12 per cent. of 1.587 lbs. is 190 lbs. of water. Water weighs 8.355 lbs. per gallon, hence 190 × 8.355 = 23 gallons of water per cubic yard of 1-3-6 concrete.

METHODS OF MEASURING AND WEIGHING.—The cement, sand and aggregate for concrete mixtures are usually measured by hand, the measuring being done either in the charging buckets or in the barrows or other receptacles used to handle the material to the charging buckets. The process is simple in either case when once the units of measurement are definitely stated. This is not always the case. Some engineers require the contractor to measure the sand and stone in the same sized barrel that the cement comes in, in which case 1 part of sand or aggregate usually means 3.5 cu. ft. Other engineers permit both heads of the barrel to be knocked out for convenience in measuring the sand and stone, in which case a barrel means 3.75 cu. ft. Still other engineers permit the cement to be measured loose in a box, then a barrel usually means from 4 to 4.5 cu. ft. Cement is shipped either in barrels or in bags and the engineer should specify definitely the volume at which he will allow the original package to be counted, and also, if cement barrels are to be used in measuring the sand and stone, he should specify what a "barrel" is to be. When the concrete is to be mixed by hand the better practice is to measure the sand and stone in bottomless boxes of the general type shown by Fig. 10 and of known volume, and then specify that a bag of cement shall be called 1 cu. ft., 0.6 cu. ft., or such other fraction of a cubic foot as the engineer may choose. The contractor then has a definite basis on which to estimate the quantity of cement required for any specified mixture. The same is true if the measuring of the sand and stone be done in barrows or in the charging bucket. The volume of the bag or barrel of cement being specified the contractor has a definite and simple problem to solve in measuring his materials.

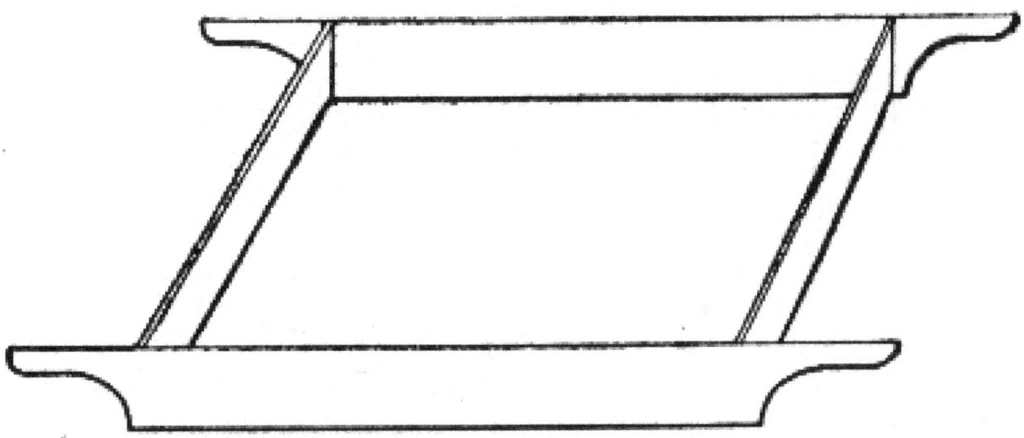

Fig. 10.—Bottomless Box for Measuring Materials in Proportioning Concrete.

To avoid uncertainty and labor in measuring the cement, sand and stone or gravel various automatic measuring devices have been designed. A continuous mixer with automatic measuring and charging mechanism is described in Chapter XIV. Figure 11 shows the Trump automatic measuring device. It consists of a series of revolving cylinders, each opening onto a "table," which revolves with the cylinders, and of a set of fixed "knives," which, as the "tables" revolve, scrape off portions of the material discharged from each cylinder onto its "table." The illustration shows a set of two cylinders; for concrete work a third cylinder is added. The three tables are set one above the other, each with its storage cylinder, and being attached to the same spindle all revolve together. For each table there is a knife with its own adjusting mechanism. These knives may be adjusted at will to vary the percentage of material scraped off.

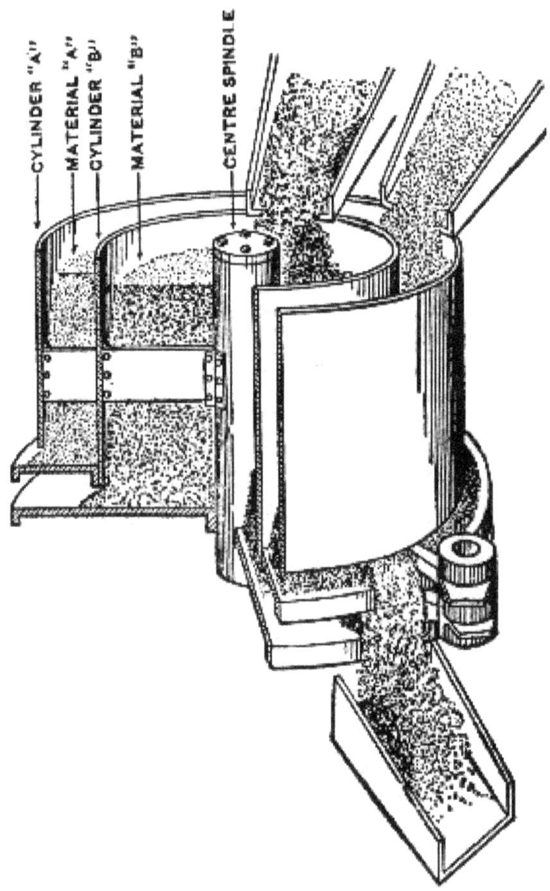

Fig. 11.—Sketch Showing Trump Automatic Measuring Device for Materials in Proportioning Concrete.

Automatic measuring devices are most used in connection with continuous mixers, but they may be easily adapted to batch mixers if desired. One point to be observed is that all of these automatic devices measure the cement loose and this must be allowed for in proportioning the mixture.

3

CONCRETE MAKING AND PLACING: METHODS AND COSTS

The making and placing of concrete by hand is divided into the following operations: (1) Loading the barrows, buckets, carts or cars used to transport the cement, sand and stone to the mixing board; (2) Transporting and dumping the material; (3) Mixing the material by turning with shovels and hoes; (4) Loading the concrete by shovels into barrows, buckets, carts or cars; (5) Transporting the concrete to place; (6) Dumping and spreading; (7) Ramming.

LOADING INTO STOCK PILES.—Stock piles should always be provided unless there is some very good reason to the contrary. They prevent stoppage of work through irregularities in the delivery of the material, and they save foreman's time in watching that the material is delivered as promptly as needed for the work immediately in hand. The location of the stock piles should be as close to the work as possible without being in the way of construction; forethought both in locating the piles and in proportioning their size to the work will save the contractor money.

The stone and sand will ordinarily be delivered in wagons or cars. If delivered in cars, effort should be made to secure delivery in flat cars when the unloading is to be done by shoveling; this is more particularly necessary for the broken stone. Stone can be shoveled from hopper bottom cars only with difficulty as compared with shoveling from flat bottom cars; the ratio is about 14 cu. yds. per day per man from hopper bottom cars as compared with 20 cu. yds. per day per man from flat bottom cars. When the cars can be unloaded through a trestle, hopper bottom cars should by all means be secured for delivering the stone. If the amount of work will justify the expense, a trestle may be built; often there is a railway embankment which can be dug away for a short distance and the track carried on stringers to make a dumping place, from which the stone can be shoveled.

Sand can be dumped directly on the ground, but broken stone unless it is very small, ¾-in. or less, should always be dumped on a well made plank floor. A good floor is made of 2-in. plank, nailed to 4×6-in. mud sills, spaced 3 ft. apart, and well bedded in the ground. Loose plank laid directly on the ground settle unevenly and thus the smooth shoveling surface which is sought is not obtained; the object of the floor is to provide an even surface, along which a square pointed shovel can be pushed; it is very difficult to force such a shovel into broken stone unless it is very fine. A spading fork is a better tool than a shovel, with which to load broken stone from piles. A man can load from 18 to 20 cu. yds. of broken stone into wheelbarrows or carts in 10 hours when shoveling from a good floor, but he can load only 12 to 14 cu. yds. per day when shoveling from a pile without such a floor. It is a common thing to see stone unloaded from cars directly onto the sloping side of a railway embankment. This makes very difficult shoveling and results in a waste of stone. Stone can usually be delivered by a steel lined chute directly to a flooring located at the foot of the embankment; coarse broken stone if given a start when cast from a shovel will slide on an iron chute having a slope as flat as 3 or 4 to 1; sand will not slide on a slope of 1½ to 1. When chuting is not practicable it will pay often to shovel the stone into buckets handled by a stiff-leg derrick rather than to unload it onto the bank. Stock piles of ample storage capacity are essential when delivery is by rail, because of the uncertainty of rail shipments. When the contractor is taking the sand and stone direct from pit and quarry by wagon it is not necessary to have large stock piles.

LOADING FROM STOCK PILES.—In loading sand into wheelbarrows or carts with shovels a man will load 20 cu. yds. per 10-hour day if he is energetic and is working under a good foreman. Under opposite conditions 15 cu. yds. per man per day is all that it is safe to count on. A man shoveling from a good floor will load 20 cu. yds. of stone per 10-hour day; this is reduced to 15 cu. yds. per day if the stone is shoveled off the ground and to 12 cu. yds. per day if in addition the management is poor. There are ordinarily in a cubic yard of concrete about 1 cu. yd. of stone and 0.4 cu. yd. of sand, so that the cost of loading the materials into barrows or carts, with wages at 15 cts. per hour and assuming 15 cu. yds. to be a day's work, would be:

1 cu. yd. stone loaded for 10 cts.

0.4 cu. yd. sand loaded for	4 cts.
Total	14 cts.

To this is to be added the cost of loading the cement. This will cost not over 2 cts. per cu. yd. of concrete; the total cost of loading concrete materials into barrows or carts, therefore, does not often exceed 16 cts. per cu. yd. of concrete.

TRANSPORTING MATERIALS TO MIXING BOARDS—Carrying the sand and stone from stock piles to mixing board in shovels should never be practiced. It takes from 100 to 150 shovelfuls of stone to make 1 cu. yd.; it, therefore, costs 50 cts. per cu. yd. to carry it 100 ft. and return empty handed, for in walking short distances the men travel very slowly—about 150 ft. per minute. It costs more to walk a half dozen paces with stone carried in shovels than to wheel it in barrows.

The most common method of transporting materials from stock piles to mixing boards is in wheelbarrows. The usual wheelbarrow load on a level plank runway is 3 bags of cement (300 lbs) or 3 cu. ft. of sand or stone. If a steep rise must be overcome to reach the mixing platform the load will be reduced to 2 bags (200 lbs.) of cement or 2 cu. ft. of sand or stone. A man wheeling a barrow travels at a rate of 200 ft. per minute, going and coming, and loses ¾ minute each trip dumping the load, fixing run planks, etc. An active man will do 20 to 25 per cent. more work than this, while a very lazy man may do 20 per cent. less. With wages at 15 cts. per hour, the cost of wheeling materials for 1 cu. yd. of concrete may be obtained by the following rule:

To a fixed cost of 4 cts. (for lost time) add 1 ct. for every 20 ft. of distance away from the stock pile if there is a steep rise in the runway, but if the runway is level, add 1 ct. for every 30 ft. distance of haul.

Since loading the barrows, as given above, was 16 cts. per cu. yd., the total fixed cost is 16 + 4 = 20 cts. per cu. yd., to which is added 1 ct. for every 20 or 30 ft. haul depending on the grade of the runway.

The preceding figures assume the use of plank runways for the wheelbarrows. These should never be omitted, and the barrows wheeled over the ground. Even a hard packed earth path in dry weather is inferior to a plank runway and when the ground is soft or muddy the loss in efficiency of the men is serious. Where the runway must rise to the mixing board, give it a slope or grade seldom steeper than 1 in 8, and if possible flatter. Make a runway on a trestle at least 18 ins. wide, so that men will be in no danger of falling. See to it, also, that the planks are so well supported that they do not spring down when walked over, for a springy plank makes hard wheeling. If the planks are so long between the "horses" or "bents" used to support them, that they spring badly, it is usually a simple matter to nail a cleat across the underside of the planks and stand an upright strut underneath to support and stiffen the plank.

When two-wheeled carts of the type shown by Fig. 12 are used the runway requires two lines of planks.

Two-wheeled carts pushed by hand have been less used for handling concrete materials than for handling concrete, but for distances from 50 to 150 ft. from stock pile to mixing board such carts are probably cheaper for transporting materials than are wheelbarrows. These carts hold generally three wheelbarrow loads and they are handled by one man practically as rapidly and easily as is a wheelbarrow.

For all distances over 50 ft. from stock pile to mixing board, it is cheaper to haul materials in one-horse dump carts than it is in wheelbarrows. A cart should be loaded in 4 minutes and dumped in about 1 minute, making 5 minutes lost time each round trip. It should travel at a speed of not less than 200 ft. per minute, although it is not unusual to see variations of 15 or 20 per cent., one way or another, from this average, depending upon the management of the work. A one-horse cart will readily carry enough stone and sand to make ½ cu. yd. of concrete, if the roads are fairly hard and level; and a horse can pull this load up a 10 per cent. (rise of 1 ft. in 10 ft.) planked roadway provided with cleats to give a foothold. If a horse, cart and driver can be hired for 30 cts. per hour, the cost of hauling the materials for 1 cu. yd. of concrete is given by the following rule:

To a fixed cost of 5 cts. (for lost time at both ends of haul) add 1 ct. for every 100 ft. of distance from stock pile to mixing board.

Fig. 12.—Two-Wheeled Ransome Cart for Hauling Concrete.

Where carts are used it is possible to locate the stock piles several hundred feet from the mixing boards without adding materially to the cost of the concrete. It is well, however, to have the stock piles in sight of the foreman at the mixing board, so as to insure promptness of delivery.

METHODS AND COST OF MIXING.—In mixing concrete by hand the materials are spread in superimposed layers on a mixing board and mixed together first dry and then with water by turning them with shovels or hoes. The number of turns, the relative arrangement of the layers, and the sequence of operations vary in practice with the notions of the engineer controlling the work. No one mode of procedure in hand mixing can, therefore, be specified for general application; the following are representative examples of practice in hand mixing:

Measure the stone in a bottomless box and spread it until its thickness in inches equals its parts by volume. Measure the sand in a bottomless box set on the stone and spread the sand evenly over the stone layer. Place the cement on the sand and spread evenly. Turn the material twice with a square pointed shovel and then turn it a third time while water is gently sprinkled on. A fourth turn is made to mix thoroughly the water and the concrete is then shoveled into barrows, giving it a fifth turn. Mr. Ernest McCullough, who gives this method, states that it is the cheapest way to mix concrete by hand and still secure a good quality of output.

In work done by Mr. H. P. Boardman the sand is measured in a bottomless box and over it is spread the cement in an even layer. The cement and sand are mixed dry with hoes, the water is added in pailfuls and the whole mixed to a uniform porridge-like consistency. Into this thin mortar all the stone for a batch is dumped, the measuring box is lifted and the mixture turned by shovels. A pair of shovelers, one on each side, is started at one end turning the material back and working toward the opposite end. A second pair of shovelers takes the turned material and turns it again. The concrete is then shoveled into the barrows by the wheelers themselves as fast as it is turned the second time. By this method a good gang of 20 to 25 men, using two boxes, will, Mr. Boardman states, mix and place 45 to 60 cu. yds. of concrete in 10 hours, depending on the wheelbarrow travel necessary. Assuming a gang of 25 men, this is a rate of 1.8 to 2.4 cu. yds. per man per 10-hour day, concrete mixed and placed.

A method somewhat similar to the one just outlined is given by Mr. O. K. Morgan. A mixing board made of ⅞-in. matched boards nailed to 2×3-in. sills is used, with a mixing box about 8 ft. long, 4 ft. wide and 10 to 12 ins. deep. This box is set alongside the mixing board and in it the cement and sand are mixed first dry and then wet; a fairly wet mortar is made. Meanwhile the stone is spread in an even layer 6 ins. thick on the mixing board and thoroughly drenched with water. The mortar from the mixing box is cast by shovels in a fairly even layer over the stone and the whole is turned two or three times with shovels, generally two turns are enough. Six men are employed; two prepare the mortar, while four get the stone in readiness, then all hands finish the operation.

The following method is given by Mr. E. Sherman Gould: Spread the sand in a thin layer on the mixing board and over it spread the cement. Mix dry with shovels, using four men, one at each corner, turning outward and then working back again. Over the dry sand and cement mixture spread the broken stone which has been previously wetted and on top of the stone apply water evenly. The water will thus percolate through the stone without splashing and evenly wet the sand and cement. Finally turn the whole, using the same number of men and the same mode of procedure as were used in dry mixing the sand and cement. Mr. Gould states that by this method the contractor should average 2 cu. yds. of mixed concrete per man per 10-hour day.

A novel method of hand mixing and an unusual record of output is described by Maj. H. M. Chittenden, U. S. A., in connection with the construction of a concrete arch bridge. The mixing was done by hand on a single board 25 ft. long and sloping slightly from one end to the other. The materials were dumped together on the upper end of the board. Sixteen men were stationed along the board, eight on each side. The first two men turned the mixture dry. Next to them stood a man who applied the water after each shovelful. The next mixers kept turning the material along and another waterman assisted in wetting it further down the board. The men at the end of the board shoveled the concrete into the carts which took it to the work. Each batch contained 18 cu. ft., or 0.644 cu. yd., and the rate of mixing was 10 cu. yds. per hour, or 6.25 cu. yds. per man per 10-hour day. The work of getting the materials properly proportioned to the mixing board is not included in this figure, but the loading of the mixed concrete is included.

It is plain from the foregoing, that specifications for hand mixing should always state the method to be followed, and particularly the number of turns necessary. If these matters are not specified the contractor has to guess at the probable requirements of the engineer. The authors have known of inspectors demanding from 6 to 9 turns of the materials when specifications were ambiguous. It should also be made clear whether or not the final shoveling into the barrows or carts constitutes a turn, and whether any subsequent shoveling of the concrete into place constitutes a turn. Inspectors and foremen have frequent disputes over these questions.

Estimates of the cost of hand mixing may usually be figured upon the number of times that the materials are to be turned by shovels. A contractor is seldom required to turn the sand and cement more than three times dry and three times wet, and then turn the mortar and stone three times. A willing workman, under a good foreman, will turn over mortar at the rate of 30 cu. yds. in 10 hours, lifting each shovelful and casting it into a pile. With wages at $1.50 and six turns, this means a cost of 5 cts. per cubic yard of mortar for each turn; as there is seldom more than 0.4 cu. yd. of mortar in a cubic yard of concrete, we have a cost of 2 cts. per cubic yard of concrete for each turn that is given the mortar. So if the mortar is given six turns before the stone is added and then the stone and mortar are mixed by three turns we have: (2 cts. × 6) + (5 cts. × 3) = 12 + 15 = 27 cts. per cubic yard for mixing concrete. In pavement foundation work two turns of the mortar followed by two turns of the mortar and stone are considered sufficient. The cost of mixing per cubic yard of concrete is then (2 cts. × 2) + (5 cts. × 2) = 4 + 10 = 14 cts. per cubic yard of concrete. One specification known to the authors, requires six turns dry and three turns wet for the mortar; under such specifications the cost of mixing the mortar would be 50 per cent. higher than in the first example assumed. On the other hand, they have seen concrete mixed for street pavement foundation with only three turns before shoveling it into place. These costs of mixing apply to work done by diligent men; easy going men will make the cost 25 to 50 per cent greater.

LOADING AND HAULING MIXED CONCRETE.—Wheelbarrows and carts are employed to haul the mixed concrete to the work. The loading of these with mixed concrete by shoveling costs less than the loading of the materials separately before mixing. While the weight is greater because of the added water the volume of the concrete is much less than that of the ingredients before mixing. Again the shoveling is done off a smooth board with the added advantage of having the material lubricated and, finally, the foreman is usually at this point to crowd the work. A good worker will load 12½ cu. yds. of concrete per 10-hour day, and with wages at $1.50 per day this would give a cost of 12 cts. per cu. yd. for loading.

Practically the same principles govern the transporting of concrete in barrows as govern the handling of the raw materials in them. The cost of wheeling concrete is practically the same as for wheeling the dry ingredients, so that the total cost of loading and wheeling may be estimated by the following rule:

To a fixed cost of 16 cts. for loading and lost time add 1 ct. for every 30 ft. of level haul.

Within a few years wheelbarrows have been supplanted to a considerable extent by hand carts of the general type shown by Fig. 12, which illustrates one made by the Ransome Concrete Machinery Co. The bowl of this cart has a capacity of 6 cu. ft. water measure. It is hung on a 1¼-in. steel axle; the wheels are 42 ins. in diameter with staggered spokes and 2-in. half oval tires. The top of the bowl is 29½ ins. from the ground. Owing to the large diameter of the wheels and the fact that no weight comes on the wheeler, as with a wheelbarrow, this cart is handled by one man about as rapidly and easily as is a wheelbarrow. It will be noted that the two ends of the bowl differ in shape; the handle is removable and can be attached to either end of the bowl. With the handle attached as shown the bowl can be inverted for discharging onto a pavement or floor; with the handle transferred to the opposite end the bowl is fitted for dumping into narrow beam or wall forms. The maximum load of wet concrete for a wheelbarrow is 2 cu. ft., and this is a heavy load and one that is seldom averaged—1 to 1½ cu. ft. is more nearly the general average. A cart of the above type will, therefore, carry from 3 to 5 wheelbarrow loads, and on good runways, which are essential, may be pushed and dumped about as rapidly as a wheelbarrow. In succeeding pages are given records of actual work with hand carts which should be studied in this connection.

Portland cement concrete can be hauled a considerable distance in a dump cart or wagon before it begins to harden; natural cement sets too quickly to permit of its being hauled far. Portland cement does not begin to set in less than 30 minutes. On a good road, with no long, steep hills a team will haul a loaded wagon at a speed of about 200 ft. per minute; it, therefore, takes 6½ minutes to travel a quarter of a mile, 13 minutes to travel half a mile, and 26 minutes to travel a mile. Portland cement concrete can, therefore, be hauled a mile before it begins to set. The cost of hauling concrete in carts is about the same as the cost of hauling the raw materials as given in a preceding section.

When hand mixing is employed in building piers, abutments, walls, etc., the concrete often has to be hoisted as well as wheeled. A gallows frame or a mast with a pulley block at the top and a team of horses can often be used in such cases as described in Chapter XII for filling cylinder piers, or in the same chapter for constructing a bridge abutment. It is also possible often to locate the mixing board on high ground, perhaps at some little distance from the forms. If this can be done, the use of derricks may be avoided as above suggested or by building a light pole trestle from the mixing board to the forms. The concrete can then be wheeled in barrows and dumped into the forms. If the mixing board can be located on ground as high as the top of the concrete structure is to be, obviously a trestle will enable the men to wheel on a level runway. Such a trestle can be built very cheaply, especially where second-hand lumber, or lumber that can be used subsequently for forms is available. A pole trestle whose bents are made entirely of round sticks cut from the forest is a very cheap structure, if a foreman knows how to throw it together and up-end the bents after they are made. One of the authors has put up such trestles for 25 cts. per lineal foot of trestle, including all labor of cutting the round timber, erecting it, and placing a plank flooring 4 ft. wide on top. The stringers and flooring plank were used later for forms, and their cost is not included. A trestle 100 ft. long can thus be built at less cost than hauling, erecting and taking down a derrick; and once the trestle is up it saves the cost of operating a derrick.

In conclusion, it should be remarked that the comparative economy for concrete work of the different methods of haulage described, does not depend wholly on the comparative transportation costs; the effect of the method of haulage on the cost of dumping and spreading costs must be considered. For example, if carts deliver the material in such form that the cost of spreading is greatly increased over what it would be were the concrete delivered in wheelbarrows, the gain made by cart haulage may be easily wiped out or even turned into loss by the extra spreading charges. These matters are considered more at length in the succeeding section.

DUMPING, SPREADING AND RAMMING.—The cost of dumping wheelbarrows and carts is included in the rules of cost already given, excepting that in some cases it is necessary to add the wages of a man at the dump who assists the cart drivers or the barrow men. Thus in dumping concrete from barrows into a deep trench or pit, it is usually advisable to dump into a galvanized iron hopper provided with an iron pipe chute. One man can readily dump all the barrows that can be filled from a concrete mixer in a day, say 150 cu. yds. At this rate of output the cost of dumping would be only 1 ct. per cu. yd., but if one man were required to dump the output of a small gang of men, say 25 cu. yds., the cost of dumping would be 6 cts. per cu. yd.

Concrete dumped through a chute requires very little work to spread it in 6-in. layers; and, in fact, concrete that can be dumped from wheelbarrows, which do not all dump in one place, can be spread very cheaply; for not more than half the pile dumped from the barrow needs to be moved, and then moved merely by pushing with a shovel. Since the spreader also rams the concrete, it is difficult to separate these two items. As nearly as the authors have been able to estimate this item of spreading "dry" concrete dumped from wheelbarrows in street paving work, the cost is 5 cts. per cu. yd. If, on the other hand, nearly all the concrete must be handled by the spreaders, as in spreading concrete dumped from carts, the cost is fully double, or 10 cts. per cu. yd. And if the spreader has to walk even 3 or 4 paces to place the concrete after shoveling it up, the cost of spreading will be 15 cts. per cu. yd. For this reason it is apparent that carts are not as economical as wheelbarrows for hauling concrete up to about 200 ft., due to the added cost of spreading material delivered by carts.

The preceding discussion of spreading is based upon the assumption that the concrete is not so wet that it will run. Obviously where concrete is made of small stones and contains an excess of water, it will run so readily as to require little or no spreading.

The cost of ramming concrete depends almost entirely upon its dryness and upon the number of cubic yards delivered to the rammers. Concrete that is mixed with very little water requires long and hard ramming to flush the water to the surface. The yardage delivered to the rammers is another factor, because if only a few men are engaged in mixing they will not be able to deliver enough concrete to keep the rammers properly busy, yet the rammers by slow though continuous pounding may be keeping up an appearance of working. Then, again, it has been noticed that the slower the concrete is delivered the more particular the average inspector becomes. Concrete made "sloppy" requires no ramming at all, and very little spading. The authors have had men do very thorough ramming of moderately dry concrete for 15 cts. per cu. yd., where the rammers had no spreading to do, the material being delivered in shovels. It is rare indeed that spreading and ramming can be made to cost more than 40 cts. per cu. yd., under the most foolish inspection, yet one instance is recorded which, because of its rarity, is worth noting: Mr. Herman Conrow is the authority for the data: 1 foreman, 9 men mixing, 1 ramming, averaged 15 cu. yds. a day, or only 1½ cu. yds. per man per day, when laying wet concrete. When laying dry concrete the same gang averaged only 8 cu. yds. a day, there being 4 men ramming. With foreman at $2 and laborers at $1.50 a day, the cost was $2.12 per cu. yd. for labor on the dry concrete as against $1.13 per cu. yd. for the wet concrete. Three turnings of the stone with a wet mortar effected a better mixture than four turnings with a dry mortar. The ramming of the wet concrete cost 10 cts. per cu. yd., whereas the ramming of the dry concrete cost 75 cts. per cu. yd. The authors think this is the highest cost on record for ramming. It is evident, however, that the men were under a poor foreman, for an output of only 15 cu. yds. per day with 10 men is very low for ordinary conditions. Moreover, the expensive amount of ramming indicates either poor management or the most foolish inspection requirements.

In conclusion it may be noted that if engineers specify a dry concrete and "thorough ramming," they would do well also to specify what the word "thorough" is to mean, using language that can be expressed in cents per cubic yard. It is a common thing, for example, to see a sewer trench specification in which one tamper is required for each two men shoveling the back-fill into the trench; and some such specific requirement should be made in a concrete specification if close estimates from reliable contractors are desired. Surely no engineer will claim that this is too unimportant a matter for consideration when it is known that ramming can easily be made to cost as high as 40 cts. per cu. yd., depending largely upon the whim of the inspector.

THE COST OF SUPERINTENDENCE.—This item is obviously dependent upon the yardage of concrete handled under one foreman and the daily wages of the foreman. If a foreman receives $3 a day and is bossing a job where only 12 cu. yds. are placed daily, we have a cost of 25 cts. per cu. yd. for superintendence. If the same foreman is handling a gang of 20 men whose output is 50 cu. yds., the superintendence item is only 6 cts. per cu. yd. If the same foreman is handling a concrete-mixing plant having a daily output of 150 cu. yds., the cost of superintendence is but 2 cts. per cu. yd. These elementary examples have been given simply because figures are more impressive than generalities, and because it is so common a sight to see money wasted by running too small a gang of men under one foreman.

Of all classes of contract work, none is more readily estimated day by day than concrete work, not only because it is usually built in regular shapes whose volumes are easily ascertained at the end of each day, but because a record of the bags, or barrels, or batches gives a ready method of computing the output of each gang. For this reason small gangs of concrete workers need no foreman at all, provided one of the workers is given command and required to keep tally of the batches. If the efficiency of a gang of 6 men were to fall off, say, 15 per cent., by virtue of having no regular non-working foreman in charge, the loss would be only $1.35 a day—a loss that would be more than counterbalanced by the saving of a foreman's wages. Indeed, the efficiency of a gang of 6 men would have to fall off 25 per cent., or more, before it would pay to put a foreman in charge. In many cases the efficiency will not fall off at all, provided the gang knows that its daily progress is being recorded, and that prompt discharge will follow laziness. Indeed, one of the authors has more than once had the efficiency increased by leaving a small gang to themselves in command of one of the workers who was required to punch a hole in a card for every batch.

To reduce the cost of superintendence there is no surer method than to work two gangs of 18 to 20 men, side by side, each gang under a separate foreman who is striving to make a better showing than his competitor. This is done with marked advantage in street paving, and could be done elsewhere oftener than it is.

In addition to the cost of a foreman in direct charge of the laborers, there is always a percentage of the cost of general superintendence and office expenses to be added. In some cases a general superintendent is put in charge of one or two foremen; and, if he is a high-salaried man, the cost of superintendence becomes a very appreciable item.

SUMMARY OF COSTS.—Having thus analyzed the costs of making and placing concrete, we can understand why it is that printed records of costs vary so greatly. Moreover, we are enabled to estimate the labor cost with far more accuracy than we can guess it; for by studying the requirements of the specifications, and the local conditions governing the placing of stock piles, mixing boards, etc., we can estimate each item with considerable accuracy· The purpose, however, has not been solely to show how to predict the labor cost, but also to indicate to contractors and their foremen some of the many possibilities of reducing the cost of work once the contract has been secured. An analysis of costs, such as above given, is the most effective way of discovering unnecessary "leaks," and of opening one's eyes to the possibilities of effecting economies in any given case.

To indicate the method of summarizing the costs of making concrete by hand, let us assume that the concrete is to be put into a deep foundation requiring wheeling a distance of 30 ft.; that the stock piles are on plank 60 ft. distant from the mixing board; that the specifications call for 6 turns of gravel concrete thoroughly rammed in 6-in. layers; and that a good sized gang of, say, 16 men (at $1.50 a day each), is to work under a foreman receiving $2.70 a day. We then have the following summary by applying the rules already given:

	Per cu. yd. concrete.
Loading sand, stone and cement	$ .17
Wheeling 60 ft. in barrows (4 + 2 cts.)	.06
Mixing concrete, 6 turns at 5 cts.	.30
Wheeling 30 ft. (4 + 1 ct.)	.05
Dumping barrows (1 man helping barrowman)	.05
Spreading and heavy ramming	.15
Total cost of labor	$.90
Foreman, at $2.70 a day	.10
Grand total	$1.00

To estimate the daily output of this gang of 16 laborers proceed thus: Divide the daily wages of all the 16 men, expressed in cents, by the labor cost of the concrete in cents, the quotient will be the cubic yards output of the gang. Thus, 2,400 ÷ 90 is 27 cu. yds., in this case.

In street paving work where no man is needed to help dump the wheelbarrows, and where it is usually possible to shovel concrete direct from the mixing board into place, and where half as much ramming as above assumed is usually satisfactory, we see that the last four labor items instead of amounting to 12 + 5 + 5 + 15, or 37 cts., amount only to one-half of the last item, one-half of 15 cts., or 7½ cts. This makes the total labor cost only 60 cts. instead of 90 cts. If we divide 2,400 cts. (the total day's wages of 16 men) by 60 cts. (the labor cost per cu. yd.), we have 40, which is the cubic yards output of the 16 men. This greater output of the 16 men reduces the cost of superintendence to 7 cts. per cu. yd.

4

MAKING AND PLACING CONCRETE BY MACHINE

The making and placing of concrete is virtually a manufacturing process. This process as performed by manual labor is discussed in the preceding chapter; it will be discussed here as it is performed by machinery. The objects sought in using machinery for making and placing concrete are: (1) The securing of a more perfectly mixed and uniform concrete, and (2) the securing of a cheaper cost of concrete in place. As in every other manufacturing process both objects cannot be obtained to the highest degree without co-ordinate and universal efficiency throughout in plant and methods. For example, the substitution of machine mixing for hand mixing will not alone ensure cheaper concrete. If all materials are delivered to the machine in wheelbarrows and if the concrete is conveyed away in wheelbarrows, the cost of making concrete even with machine mixers is high. On the other hand, where the materials are fed from bins by gravity into the mixer and when the mixed concrete is hauled away in cars, the cost of making the concrete may be very low. Making and placing concrete by machinery involves not one but several mechanical operations working in conjunction—in a word, a concrete making plant is required.

The mechanical equipment of a concrete making plant has four duties to perform. (1) It has to transport the raw materials from the cars or boats or pits and place them in the stock piles or storage bins; (2) it has to take the raw materials from stock and charge them to the mixer; (3) it has to mix the raw materials into concrete and discharge the mixture into transportable vehicles; and (4) it has to transport these vehicles from the mixer to the work and discharge them. As all these operations are interrelated component parts of one great process, it is plain why one operation cannot lag without causing all the other operations to slow up.

The mechanical devices which may be used for each of these operations are various, and they may be combined in various ways to make the complete train of machinery necessary to the complete process. In this chapter we shall describe the character and qualities of each type of devices separately. The practicable ways of combining them to form a complete concrete making plant are best illustrated by descriptions and records of work of actual plants, and such descriptions and records for each class of structure considered in this book are given in the following chapters and may be found by consulting the index. In describing the various machines and devices we have made one classification for those used in handling raw materials and mixed concrete, for the reason that nearly all of them are suitable for either purpose.

UNLOADING WITH GRAB BUCKETS.—The orange-peel or clam-shell bucket is an excellent device for unloading sand or stone from cars or barges. The cost of unloading, including cleaning up the portions not reached by the bucket, is not more than from 2 to 5 cts. per cu. yd. A grab bucket of either of these types can be applied to any derrick. In unloading broken stone from barges at Ossining, N. Y., a Hayward clam-shell on a stiff-leg derrick unloaded 100 cu. yds. of broken stone per day from barge into wagons, with one engineman and one helper. In addition to the bucket work there was 24 hours' labor cleaning on each 500-cu. yd. barge load. The labor cost of unloading a 500-cu. yd. barge was as follows:

	Per Cu. Yd.
One engineman, at $2.50	2.5 cts.
One helper, at $1.50	1.5 cts.
Labor cleaning, at $1.50	0.7 cts.
Total cost per cubic yard	4.7 cts.

INCLINES.—Inclines to reach the tops of mixer and storage bins and the level of concrete work can be operated on about the following grades: For teams hauling wagons or cars, 2 per cent. maximum grade. A single heavy team will haul a 5-cu. yd. car, with ordinary bearings, weighing 2½ tons empty and 12 tons loaded, with ease on a 1½ per cent. grade, and with some difficulty on a 2 per cent. grade. A locomotive will handle cars on a grade of from 4 to 5 per cent. For team haulage 20-lb. rails may be used, and for locomotives 30-lb. rails. Grades steeper than about 5 per cent. require cable haulage.

TRESTLE AND CAR PLANTS.—Trestle and car plants for handling both concrete materials and mixed concrete have a wide range of application and numerous examples of such plants are described in succeeding chapters, and are noted in the index at the end of the book. The following estimates of the cost of a trestle and car plant are given by Mr Wm. G. Fargo. The work is assumed to cover an area of 100×200 ft. and to have three-fourths of its bulk below the economical elevation of the mixer which stands within 50 ft. of the near side of the work. If the work is under 3,000 cu. yds. in bulk and there is a reasonable time limit for completion one mixer of 200 cu. yds. capacity per 10-hour day is assumed to be sufficient. The items of car plant cost will be about as follows:

150 ft. trestle, at $1.50	$225
5 split switches with spring bridles, at $18	90
2 iron turntables, at $30	60
3⅔ cu. yd. steel cars with roller bearings	190
Total	$565

The trestle assumed is double 24-in. gage track, 6 ft. on centers; stringers 6×8 ins.×22 to 24 ft.; ties 2×6 ins., 2½ ft. on centers; running boards 2×12 ins. for each track, and 12-lb. rails; trestle legs, average length 30 ft., of green poles at 5 cts. per foot. This outfit with repairs and renewals amounting to 10 per cent., is considered good for five season's work and the timber work for several jobs if not too far apart. The yearly rental on the basis of five seasons' work would be $124.30, or $1 per working day for a season of five months. Three cars delivering ½ cu. yd. batches can deliver 200 cu. yds. of concrete, an average of 100 ft. from the mixer in 10 hours. Five men, including a man tending switches and turntable and one man to help dump, can operate the plant. With wages at $1.75 per day the labor cost of handling 200 cu. yds. of concrete would be 4⅛cts. per cu. yd.

CABLEWAYS.—Cableways arranged to span the work and if the area is wide to travel across the work at right angles to the span will handle concrete, concrete materials, forms, steel and supplies with great economy. They are particularly suitable for bridge and dam work, filter and reservoir work, building foundations and low buildings. The arrangement of a cableway plant for bridge work is described in Chapter XVII. A cableway of 800 ft. clear span on fixed towers 45 ft. high will cost complete from $4,500 to $5,000, and will handle 200 cu. yds. of concrete per 10-hour day. To put the cableway on traveling towers will cost about $1,000 more. In constructing the Pittsburg filtration work four traveling cableways from 250 to 600 ft. span were used. The towers were from 50 to 60 ft. in height and each traveled on a 5-rail track. The cableways were self-propelling. With conditions favorable each cableway delivered 300 cu. yds. of concrete per day. A cableway plant for heavy fortification work is described in Chapter XI.

BELT CONVEYORS.—Belt conveyors may be used successfully for handling both concrete materials and mixed concrete. For handling wet concrete the slope must be quite flat, and the belt must be provided with some means of cleaning off the sticky mortar paste. In several cases rotating brushes stationed at the end of the belt, where it turns over the tail pulley, have worked successfully; these brushes sweep the belt clean. Except for the cleaning device the ordinary arrangement of belt conveyor for dry materials serves for concrete.

In constructing a large gas works at Astoria, Long Island, near New York city, belt conveyors were used to handle both the sand, gravel and cement bags and the mixed concrete. The belt for handling sand and gravel is shown by Fig. 13. A derrick operating a clam-shell unloaded the sand and gravel into a small hopper, discharging into dump cars operated by a "dinky" up an incline, passing over sand and gravel storage bins. A 20-in. belt conveyor ran horizontally 105 ft. under the bins, then up an incline of 3.4 ft. in 125 ft. to feeding hoppers over the mixers. This conveyor received alternately sand and gravel by chute from the storage bins and bags of cement loaded by hand, and carried them to the feeding bins and mixer platform. The speed of the belt was 350 ft. per minute, and it required 6 h.p. to operate it when carrying 100 tons per hour. The mixing was done in two Smith mixers, which turned out 70 cu. yds. or 35 cu. yds. each per hour The mixed concrete was delivered onto a 50-ft. 24-in. belt conveyor traveling at a speed of 400 ft. per minute and dumping through a chute into cars. Only 1 h.p. was required to run the concrete conveyor. A rotating brush was used to keep the belt clean at the dumping end. It will be noted that only a small amount of power is required for operation.

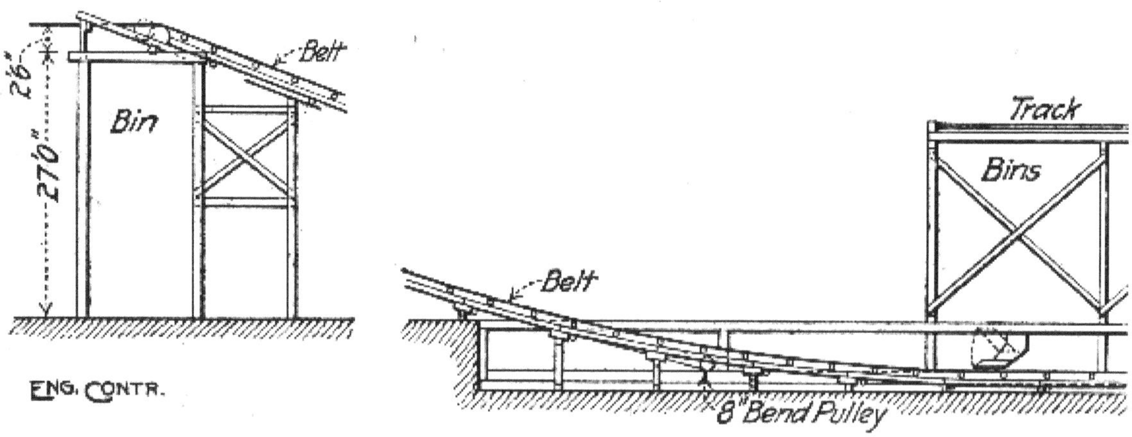

Fig. 13.—Belt Conveyor Transporting Sand and Gravel.

CHUTES.—Chutes of wood or iron are among the simplest and most efficient means of moving the cement, sand and stone and the mixed concrete when the ground levels permit such devices.

Bags of cement if given a start in casting will slide down a steel or very smooth wooden chute with a slope of 1 ft. in 5 or 6 ft. A wooden trough 12 ins. deep and 24 ins. wide with boards dressed on the inside may be used. When the inclination is steep and the fall is great, some device is necessary to diminish the velocity of descent; the following is an example of such a device which was successfully employed in a chute of the above dimensions, 400 ft. long and having a drop of 110 ft. This chute had a maximum inclination of 45° and its lower end curved to a horizontal tangent, running into the storehouse. Near the bottom of the chute a horizontal strip was nailed across the upper edges and to it was nailed the upper end of a 20 ft., 1×12-in. board, the lower end of which rested on the bottom of the chute. Several pieces of timber spiked to the upper side loaded the lower end of this board. The cement bag in descending wedged itself into the angle between the chute and the board and lifted the latter, the spring of the board and the weight at the lower end offering enough resistance to cut down the velocity. After the chute had been in use for some time and had worn smooth it was found necessary to add two more brakes to check the bags.

Broken stone will slide down a steel or steel lined chute with a slope of 1 in 3 or 4 ft. if given a start in casting. Damp sand will not slide down a chute with a slope of 1½ in 1.

A wet cement grout will flow down a smooth plank chute, with a slope of 1 in 4 ft., and wet concrete will move on the same slope; comparatively dry concrete requires a slope of nearly 1 in 1, or 45°, to secure free movement. Mr. W. J. Douglas gives the following examples of conveying concrete by chute, prefaced by the statement that his experience indicates that concrete can thus be conveyed considerable distances without material injury if proper precautions are taken.

In the first case a semi-circular steel trough about 2 ft. wide and 1 ft. deep and 15 ft. long set on a slope of 45° was used. A lift gate of sheet steel was set in the chute about 2 ft. from the upper end. The concrete was allowed to accumulate behind this gate until a wheelbarrow load was had, when the batch was let loose by lifting the gate and was discharged into barrows at the bottom. In another case a vertical chute 15 ft. long, consisting of a 15-in. square box with a canvas end, was used. The concrete was dumped into the chute in batches of about 8 cu. ft.; two men at the bottom "cut down" the pile with hoes to keep it from coning and causing separation of the stone. In a third case a continuous mixer fed into a sheet iron lined rectangular chute about 2½ ft. wide and 1 ft. deep, with a vertical drop of 60 ft. on a slope of 1 in 1, or 45°. A gate was fixed in the chute 2 ft. from the top and at the bottom the chute fed into a pyramidal hopper 3 ft. square at the top, 1 ft. square at the bottom and 4½ ft. deep. This hopper was provided with a bottom gate and was set on legs so that its top was about 10 ft. above ground. As the concrete filled in the hopper was raised and the chute cut off. The hopper was kept full all the time and was discharged by bottom gate and spout into wheelbarrows. In a fourth case the apparatus shown by the sketch, Fig. 14, was used. The continuous mixer discharged onto an 18-in. rubber conveyor belt on conical rollers and 18 ft. long. The inner end of the conveyor frame was carried on the ground at the edge of the pit and the outer end was supported by ropes from the top of a gallows frame standing on the pit bottom. The belt discharged over end into a vertical steel chute 12 ins. in diameter and 8 ft. long; this chute was fastened to the conveyor frame. Encircling and overlapping the 12-in. chute was a second slightly larger chute suspended by means of two ropes from the gallows frame. The bottom of this second chute was kept about 6 ins. below the top edges of a pyramidal hopper like the one described above. In operation the chutes and the hopper were kept filled with concrete so that the only drop of the concrete was 3 ft. from the conveyor belt into the topmost chute.

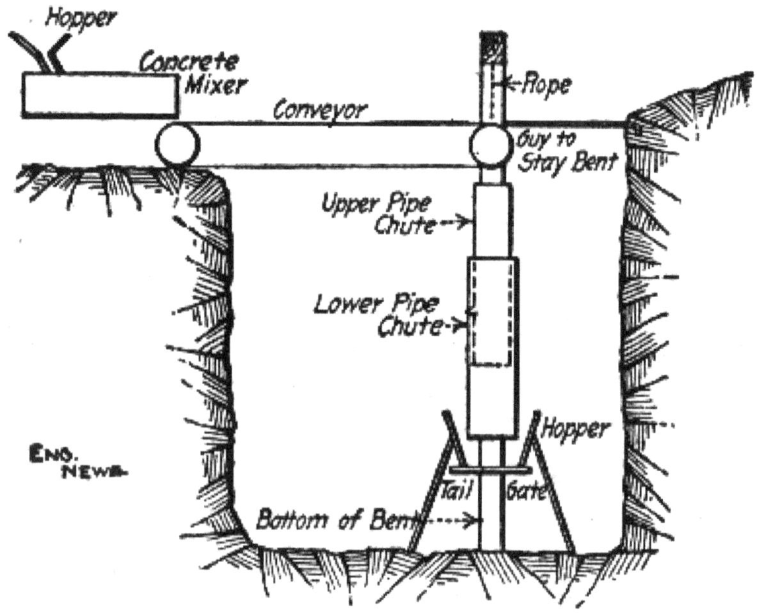

Fig. 14.—Belt Conveyor and Chute for Handling Concrete.

Concrete may be handled in long flat chutes by stationing men along the chute with shovels which they work like paddles to keep the mixture moving. In one case concrete was so handled in a chute 200 ft. long having a slope of 1 in 10 ft. The chute was a V-shaped trough made of 1×12-in. boards in sections 16 ft. long. The men paddling were stationed 10 ft. apart, so that with wages at $1.50 per day the cost would be 1½ cts. per cu. yd. for every 10 ft. the concrete was conveyed. In connection with this particular work we are informed that a Eureka continuous mixer was used. The gravel was dumped near the mixer and a team hitched to a drag scraper delivered the gravel alongside the mixer. Four men shoveled the gravel into the measuring hopper, but only two men worked at a time, shoveling for a period of 15 minutes and then resting for a corresponding period while the other two men worked. In this manner the four men shoveled enough gravel to make 100 cu. yds. of concrete per day. A fifth man opened the cement bags and kept the cement hopper filled.

METHODS OF CHARGING MIXERS.—By charging is meant the process of delivering raw materials from stock into the mixer. Several methods are practiced and will be considered in the following order: (1) By gravity from overhead bins; (2) by wheelbarrow or hand cart (a) to charging chute and (b) to elevating charging hoppers; (3) by charging cars operated by cable or other means; (4) by shoveling directly into mixer; (5) by derricks or other hoists.

Charging by Gravity from Overhead Bins.—Chuting the sand and stone from overhead bins to the charging hopper is a simple, rapid and economical method of charging mixers. The bottoms of the bins should always be high enough above the charging floor to give ample head room for men to move about erect, and the length of chute may be anything reasonable more than this that conditions such as the side hill delivery of material may necessitate. When the mixer is located to one side of the bins the slope of the chute will have to be watched. Broken stone or pebbles will move on a comparatively flat slope but sand, particularly if damp, requires a steep chute. The measuring hopper is best kept entirely independent of the mixer so that it can be filled with a new charge while the mixer is turning and discharging the preceding batch. One man can attend the sand and cement chutes if they be conveniently arranged, and one man can open and empty the cement bags if they be stacked close at hand. A third man will level off the sand and stone in the measuring hopper and help in the chuting. A gang of this size will easily measure up a charge every 2 minutes when no delays occur.

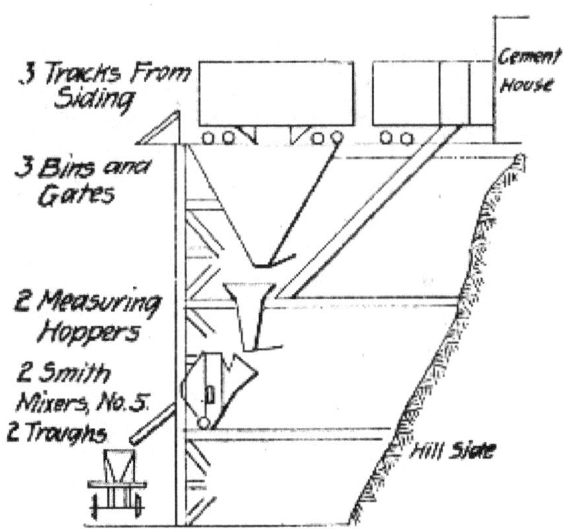

Fig. 15.—Side Hill Mixing Plant.

A number of plants charging by gravity from overhead bins are described in succeeding chapters and are referenced in the index. As a general example a side hill plant of conventional construction is shown by Fig. 15. The trestle work was made of 12×12-in. timbers and was approximately 40 ft. in height. Three tracks occupy the top platform. Under each track was a material bin; one on each side for gravel and a middle bin for sand. The sand bin was divided by a partition into two compartments. These bins discharged into two measuring hoppers one gravel bin and one compartment of the sand bin into each hopper. Two cement chutes from the top platform provided for the delivery of the cement to the mixers, either directly from cars or from the cement storage house. The mixing was done in two Smith No. 5 mixers, one under each measuring hopper, and these mixers discharged by chutes into buckets on flat cars. Thus the concrete materials brought directly from a siding in car load lots to the top of the platform were handled entirely by gravity to the cars delivering the mixed concrete to the work. The gang operating the mixing plant, with the wages paid, was composed as follows: 1 foreman and engineer at $3 per day, 1 fireman at $2 per day and 15 laborers at $1.50 per day. With this gang the two mixers turned out 400 cu. yds. of concrete per day and, frequently, 800 cu. yds. in 24 hours. Taking these figures the labor cost from raw materials in cars on the platform to mixed concrete in cars on the delivery track was as follows:

1 foreman and engineer at $3	$ 3.00
1 fireman at $2	2.00
15 laborers at $1.50	22.50
Total labor	$27.50

Assuming 400 cu. yds. output, this gives a cost of $27.50 ÷ 400 = 6.875 cts. per cu. yd.

Charging with Wheelbarrows.—The economics of wheelbarrow haulage are discussed in some detail in Chapter III. For machine mixer work the problem of loading, transporting and dumping is complicated by the greater rapidity with which the mixing is done and by the necessity, usually, of using inclines to reach the charging hopper level. The incline cuts down the output of the wheelers or in other words makes necessary a larger gang to handle the same amount of material. Conditions being the same, the height of the charging chute of the mixer determines the height of incline and the size of the charging gang, so that a mixer with a high charging level costs more to charge with wheelbarrows than does one with a low charging level. Exact figures of the increased cost of a few feet extra elevation of the wheelbarrow incline are not available, but some idea may be had from a brief calculation. The materials for a cubic yard of concrete will weigh about 3,700 lbs., so that to raise the materials for 100 cu. yds. of concrete, including weight of barrows, 1 ft. calls for about 400,000 ft. lbs. of work. A man will do about 800,000 ft. lbs. of useful work in a day, so that each foot of additional height of incline means an additional half-day's work for one man.

Wheeling to elevating charging hoppers obviates the use of inclines. Figure 19 shows a mixer equipped with such a hopper, and the arrangement provided for other makes of mixer is much similar. When the hopper is lowered ready to receive its load its top edge over which the wheelbarrows are dumped is from 12 to 14 ins. above ground level. The wheeling is all done on the level. The elevating bucket is operated by the mixer engine and is usually detachable. Where mixers have to be moved frequently, requiring the erection and moving of the incline each time, an elevating charging hopper is particularly useful; it can be hoisted clear of the ground and moved with the mixer, so that it is ready to use the moment that the mixer is set at its new station.

While the ordinary wheelbarrow is generally used for charging, better work can be done under some conditions by using special charging barrows of larger capacity and dumping from the end and ahead of the wheel. Two forms of charging barrow are shown by Figs. 16 and 17. The Acme barrow will hold 4 cu. ft. and the Ransome barrow is made in 3 to 6 cu. ft. capacities. Where inclines are necessary these barrows can often be hauled up the incline by power. A sprocket chain in the plane of the incline and operated by the mixer engine is an excellent arrangement. A prong riveted to the rear face of the barrow and projecting downward is "caught into" the chain, which pulls the barrow to the top, the man following to dump and return for another load.

Fig. 16.—
Forward Dump Charging Barrow, Sterling Wheelbarrow Co.

Fig. 17.—Forward Dump Charging
Barrow, Ransome Concrete Machinery Co.

Charging with Cars.—Cars moved by cable, team or hand are a particularly economic charging device when the mixer is located a little distance from the stock piles or bins. Either separate cars for cement, sand and stone, each holding the proper amount of its material for a batch, can be used, or a single car containing enough of all three materials for a batch. The last arrangement is ordinarily more economical in time and labor, and in plant required. In either case the car serves as the measuring hopper, there being no further proportioning of the materials after they have been loaded into the car, and it must be arranged for measuring. Usually all that is necessary, where one car is used, is to mark the levels on the sides to which it is to be filled with sand and then stone; the car is run to the sand stock and filled to the level marked for sand and then to the stone stock and filled to the level marked for stone. The cement may be added to the charge either before or after it is run to the mixer as convenience in storing the cement stock dictates. Instead of having marks to show the proper proportions of sand and stone, the car is sometimes divided into two compartments, one for each material and each holding the proper proportion of its material when level full. This arrangement makes proper proportioning somewhat more certain, since the men charging the car cannot over-run the marks. In case separate cars are used for each material, they are simply filled level full or to mark, and dumped in succession into the feeding hopper. Trestle and car plant construction and costs are given in a preceding section.

Charging by Shoveling.—Charging by shoveling directly into the mixer is seldom practiced except in street work with continuous mixers or in charging gravity mixers of the trough type. Shoveling is not an economic method of handling materials where the work involves carrying in shovels, and it is only in a few classes of concrete work or in isolated, exceptional cases that charging with shovels does not involve carrying. The amount of material that men will load with shovels is given in Chapter III, and the reader who wishes a full discussion of the subject is referred to Gillette and Hauer, "Earth Excavation and Embankments; Methods and Cost."

In charging continuous mixers with shovels the usual practice for mixers without automatic feed devices is to work from a continuous stock pile of sand, stone and cement spread in layers in the proper proportions. The shoveling is done in such a manner that each shovelful contains a mixture of cement, sand and stone, and so that the rate of delivery to the mixer is as uniform as possible. In charging mixers having automatic feed devices the sand and stone are simply shoveled into the sand and stone hoppers, whence they are fed automatically to the mixer. In charging gravity mixers by shoveling the method is essentially the same; the cement, sand and stone properly proportioned are spread in layers on the shoveling board at the head of the mixer and the mixture then shoveled into the mixer. In both of these cases mixing is performed to a certain extent by the shoveling, and in both the provision of the combination stock pile from which the men work involves labor which comes within the meaning of the term charging as we have used it here. Examples of street work in which the mixers were charged by shoveling are given in Chapter XIV.

Charging with Derricks.—When the stock piles are located close to the mixer and the plant is fixed or is not frequently moved derricks can be used economically for charging, particularly if the mixer be elevated so that inclines become expensive. The following mode of operation will be found to work well: Set the derrick so that its boom "covers" the sand and stone piles and the mixer, and provide it with three buckets so that there will always be one bucket at the stone pile and another at the sand pile while the third is being handled. The derrick swinging from the mixer, where it has discharged a bucket, drops the empty bucket at the stone pile and picks up the bucket standing there, which has received its proper charge of stone, and swings it to the sand pile and drops it to get its charge of sand. Here it picks up the bucket standing at the sand pile and which has its charges of both stone and sand, and swings it to the mixer. By this arrangement the work of the derrick and of the men filling the buckets is practically continuous. The buckets can be provided with marks on the inside to show the proper points to which to fill the stone and the sand or a partition may be riveted in making a compartment for sand and another for stone. A special charging-bucket that is arranged with a wheel and detachable handles which permit it to be handled like a wheelbarrow is shown by Fig. 18. This bucket can be used to advantage where the stock piles are too far from the mixer for the derrick to reach both, the bucket being loaded and wheeled to within reach of the derrick.

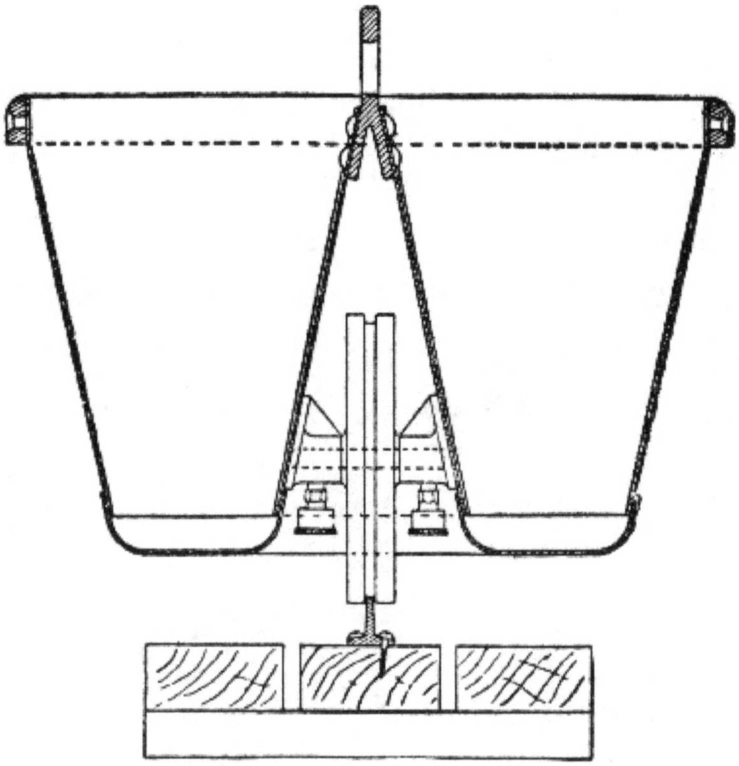

Fig. 18.—Charging Bucket With Wheel and Detachable Handle.

TYPES OF MIXERS.—There are two types of concrete mixing machines or concrete mixers as they are more commonly called: (1) Batch mixers and (2) continuous mixers. In mixers of the first type a charge of cement, sand, aggregate and water is put into the machine which mixes and discharges the batch before taking in another charge; charging, mixing and discharging is done in batches. In continuous mixers the cement sand, stone and water are charged into the machine in a continuous stream and the mixed concrete is discharged in another continuous stream. While all concrete mixers are either batch or continuous mixers, it is common practice because of their distinctive character to separate gravity mixers, whether batch or continuous, into a third type. In gravity mixers the concrete materials are made to mingle by falling through specially constructed troughs, or tubes, or hoppers. We shall describe mixers in this chapter as (1) batch mixers, (2) continuous mixers, and (3) gravity mixers. No attempt will be made, however, to describe all or even all the leading mixers of each type; a representative mixer or two of each type will be described, enough to give an indication of the range of practice, and the reader referred to manufacturers' literature for further information.

Batch Mixers.—Batch mixers are made in two principal forms which may be designated as tilting and non-tilting mixers. In the first form the mixer drum is tilted as one would tilt a bucket of water to discharge the batch. In non-tilting mixers the mixer drum remains in one position, the batch being discharged by special mechanism which dips it out a portion at a time. In both forms the charge is put into the mixer as a unit and kept confined as a unit during the time of mixing, which may be any period wished by the operator.

Fig. 19.—Chicago Improved Cube Concrete Mixer with Elevating Charging Hopper.

Chicago Improved Cube Tilting Mixer.—Figure 19 shows the improved cube mixer made by the Municipal Engineering & Contracting Co., Chicago, Ill. The drum consists of a cubical box with rounded corners and edges. This box has hollow gudgeons at two diagonally opposite corners and these gudgeons are open as shown to provide for charging and discharging. The box is rotated by gears meshing with a circumferential rack midway between gudgeons and another set of gears operate to tilt the mixer. The inside of the box is smooth, there being no deflectors, as its shape is such as to fold the batch repeatedly and thus accomplish the mixing.

Fig. 20.—Ransome Concrete Mixer.

Ransome Non-Tilting Mixer.—Figure 20 shows a representative non-tilting mixer made by the Ransome Concrete Machinery Co., Dunellen, N. J. It consists of a cylindrical drum riding on rollers and rotated by a train of gears meshing with circumferential racks on the drum. The drum has a circular opening at each end; a charging chute enters one opening and a tilting discharge chute may be thrown into or out of the opposite opening. The cylindrical shell of the drum is provided inside with steel plate deflectors, which plow through and pick up and drop the concrete mixture as the drum revolves. The shape and arrangement of the deflectors are such that the batch is shifted back and forth axially across the mixer. To discharge the batch the discharge chute is tilted so that its end projects into the mixer, in which position the material picked up by the deflectors drops back onto the chute and runs out. The discharge chute being independent of the mixing drum it can be thrown into and out of discharge position at will without stopping the rotation of the drum, and so can discharge any part or all of the batch at once. The top edge of the charging chute ranges from 30½ to 38 ins. in height above the top of the frame, varying with the size of the mixer.

Fig. 21.—Smith Concrete Mixer.

Smith Tilting Mixer.—Figure 21 shows a tilting mixer, known as the Smith mixer, made by the Contractors' Supply & Equipment Co., Chicago, Ill. The drum consists of two truncated cones with their large ends fastened together and their small ends open for receiving the charge and discharge of the batch. The drum is operated by a train of gears meshing into a rack at mid-length where the cones join. In addition there is another set of gears which tilt the drum to make the concrete flow out of the discharge end. The inside of the drum is provided with steel plate deflectors, which plow through and pick and drop the concrete mixture shifting it back and forth axially in the process.

Continuous Mixers.—Continuous mixers are those in which the cement, sand and stone are fed to the charging hopper in a continuous stream and the mixed concrete is discharged in another continuous stream. They are built in two principal forms. In one form the cement, sand and stone properly proportioned are shoveled directly into the mixing drum. In the other form these materials are dumped into separate charging hoppers and are automatically fed into the mixing drum in any relative proportions desired. One form of continuous mixer with automatic feed is described in the succeeding paragraph and another form is described in Chapter XIV. The continuous mixer without automatic feed consists simply of a trough with a rotating paddle shaft and its driving mechanism. The charging, the mixing and the discharging are done in what is virtually a succession of very small batches.

Fig. 22.—Eureka Automatic Feed Continuous Mixer.

Eureka Automatic Feed Mixer.—Figure 22 shows the construction of the continuous mixer built by the Eureka Machine Co., Lansing, Mich. The cement bin and feeder is the small one in the foreground. There is a pocketed cylinder revolving between concave plates, opening into the hopper above, from which the pockets in the feeder are filled, and discharging directly into the mixing trough below. Back of this is shown the feeder for sand or gravel up to 2-in. screen size. This is a pocketed cylinder similar to that used in the cement feeder, except that it is larger, and instead of being provided on the discharge side with a concave plate, is surmounted by a roller, held by springs. This serves to cut off the excessive flow of material, but provides sufficient flexibility to allow the rough coarse material to be fed through the machine without its catching. The feeder for crushed stone is a similar construction on larger lines, to handle material up to 3-in. size. These several feeders can be set to give any desired mixture.

On any material fit to be used in concrete, they will measure with an error of less than 5 per cent., an agitator being provided in the sand bin to prevent damp sand from bridging over the feeder, and preventing its action. The mixer consists of a trough, with a square shaft, on which are mounted 37 mixing paddles, which are slipped on in rotation, so as to form practically a continuous conveyor, but as each paddle is distinct, and is shaped like the mold board of a plow, the material, as it passes from one to the next, is turned over and stirred. Water is sprayed into the mass at the center of the trough. The result is a dry mix, followed by a wet mix. The mixing trough is made of heavy gage steel, well reinforced, and practically indestructible. To take care of the discharge of material while changing wheelbarrows, a hood is provided on the discharge end of the machine, which can be lowered, and will hold about a wheelbarrow load.

Gravity Mixers.—Gravity mixers are constructed in two general forms. The first form is a trough whose bottom or sides or both are provided with pegs, deflectors or other devices for giving the material a zig-zag motion as it flows down the trough. The second form consists of a series of hoppers set one above the other so that the batch is spilled from one into the next and is thus mixed.

The chief advantage claimed for gravity mixers is that no power is required to operate them. This is obviously so only in the sense that gravity mixers have no power-operated moving mechanism, and the fact should not be overestimated. The cost of power used in the actual performance of mixing is a very small item. The distance between feed and discharge levels is always greater for gravity mixers than for machine mixers, and the power required to raise the concrete materials the excess height may easily be greater than the power required to operate a machine mixer. On the other hand the simplicity of the gravity mixer insures low maintenance costs.

Gilbreth Trough Mixer.—Figure 23 shows the construction of one of the best known makes of gravity mixers of the trough form. In operation the cement, sand and stone in the proper proportions are spread in superimposed layers on a shoveling board at hopper level and are then shoveled as evenly as possible into the hopper. From the hopper the materials flow down the trough, receiving the water about half way down, and are mixed by being cut and turned by the pins and deflectors. The trough of the mixer is about 10 ft. long.

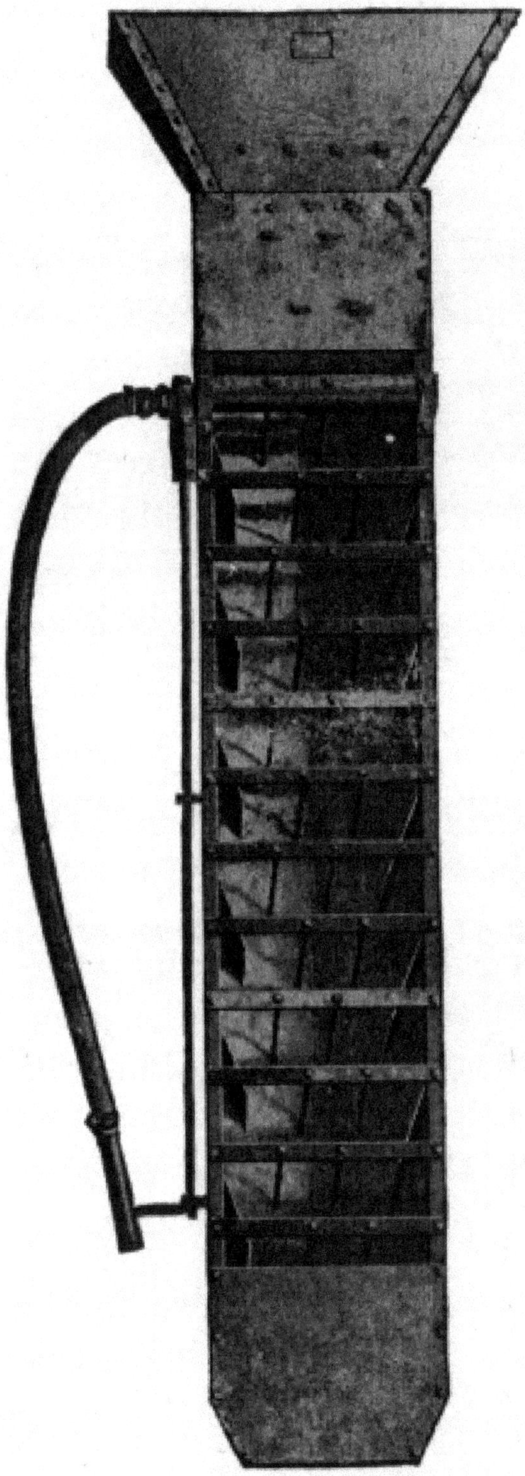

Fig. 23.—Gilbreth Gravity Mixer, Trough Form.

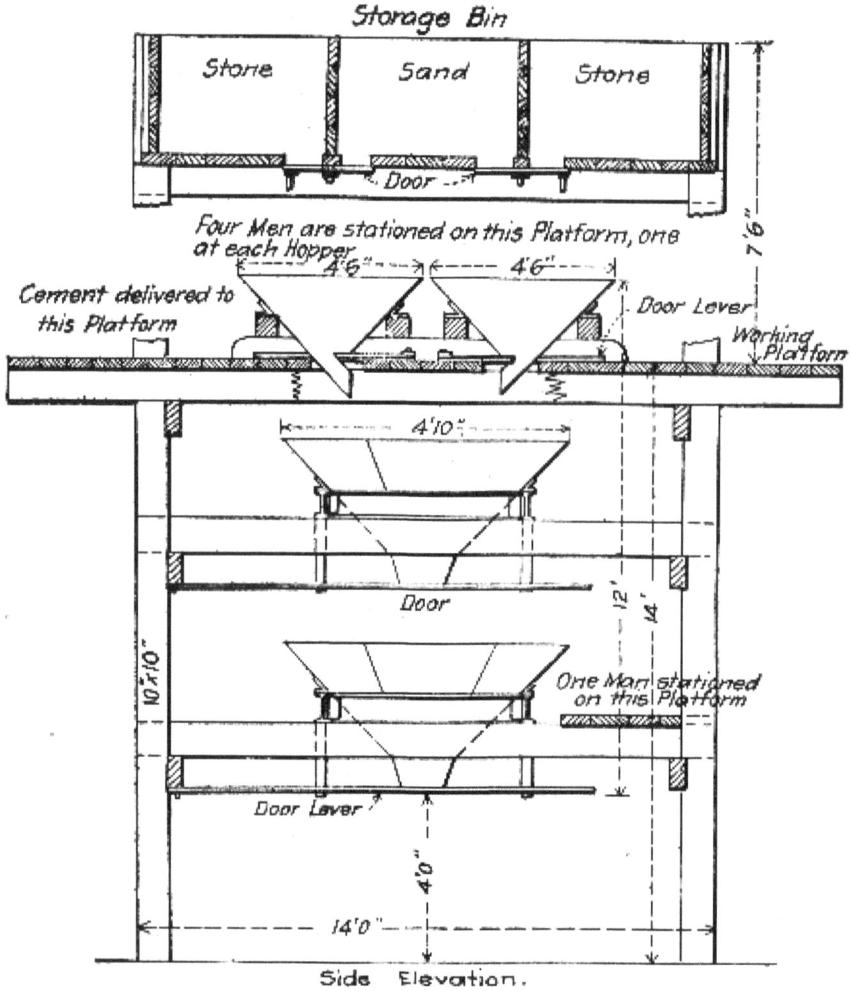

Fig. 24.—Hains Gravity Mixer, Fixed Hopper Form.

Hains Gravity Mixer.—The form of gravity mixer made by the Hains Concrete Mixer Co., Washington, D. C., is shown by Figs. 24 and 25. The charge passes through the hoppers in succession. Considering first the stationary plant, shown by Fig. 24, the four hoppers at the top have a combined capacity of one of the lower hoppers. Each top hopper is charged with cement, sand and stone in the order named and in the proper proportions. Water is then dashed over the tops of the filled hoppers and they are dumped simultaneously into the hopper next below. This hopper is then discharged into the next and so on to the bottom. Meanwhile the four top hoppers have been charged with materials for another batch. It will be observed that (1) the concrete is mixed in separate batches and (2) the ingredients making a batch are accurately proportioned and begin to be mixed for the whole batch at once. The best arrangement is to have the top of the hopper tower carry sand and stone bins which chute directly into the top hoppers. In the telescopic mixer shown by Fig. 25 the purpose has been to provide a mixer which, hung from a derrick or cableway, will receive a charge of raw materials at stock pile and deliver a batch of mixed concrete to the work, the operation of mixing being performed during the hoist to the work. By providing two mixers so that one can be charged while the other is being hoisted continuous operation is secured. The following are records of operation of stationary gravity mixers of this type.

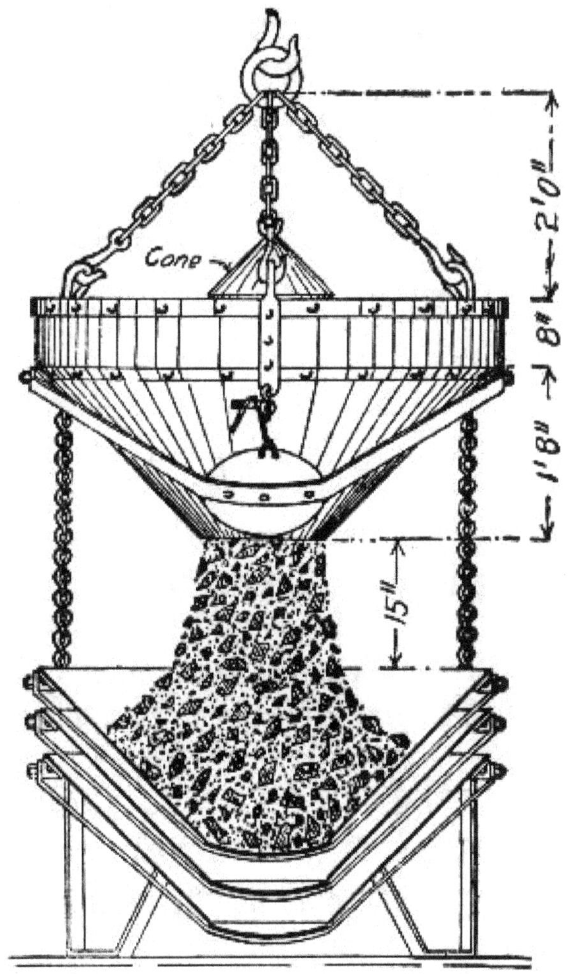

Fig. 25.—Hains Gravity Mixer, Telescoping Hopper Form.

In building a dock at Baltimore, Md., a plant consisting of two large hoppers and four charging hoppers with sand and stone bins above was used. One man at each large conical hopper tending the gates and two men charging the four pyramidal hoppers composed the mixer gang. A scow load of sand and another of stone were moored alongside the work and a clam-shell bucket dredge loaded the material from these barges into the mixer bins. Each batch was 25 cu. ft. of 1-2-5 concrete rammed in place. The men at the upper hoppers would empty a sack of cement in each, and then by opening gates in the bottom of the bins above, allow the necessary amounts of sand and stone to flow in, marks having been previously made on the sides of the hoppers to show the correct proportion of each of the ingredients. The amount of water found by experience to be necessary, would then be dashed into the hoppers, and the charges allowed to run into the first cone hopper below. Refilling would begin at the top while the men were caring for the first charge in the lower hoppers. The process was thus continuous. The concrete was chuted directly into place from the bottom hopper. The record of output was 110 batches per 10-hour day. Wages of common labor were $1.50 per day. The labor cost per cubic yard of concrete in place was 35 cts.

In constructing the Cedar Grove reservoir at Newark, N. J., a Hains mixer made the following records of output:

	Cu. yds.
Best output per 10-hour day	403
Average daily output for best month	302
Average daily output for whole job	225

The stone, sand and cement were all raised by bucket elevators to the top of the high wooden tower that supported the bins and mixer. There were 10 men operating the mixer so that (exclusive of power, interest and depreciation) the labor cost of mixing averaged only 7 cts. per cu. yd.; during one month it was as low as 5 cts. per cu. yd. This does not include delivering the materials to the men at the mixer, nor does it include conveying the concrete away and placing it. The work was done by contract.

OUTPUT OF MIXERS.—With a good mixer the output depends upon the methods of conveying the materials to and from the mixer. Most makers of mixers publish capacities of their machines in batches or cubic yards output per hour; these figures may generally be taken as stating nearly the maximum output possible. Considering batch mixers, as being the type most commonly used, it may be assumed that where the work is well organized and no delay occurs in delivering the materials to the mixer that a batch every 2 minutes, or 300 batches in 10 hours, will be averaged, and there are a few records of a batch every 1½ minutes.

To illustrate to how great an extent the output of a mixer depends on the methods adopted in handling the materials to and from the mixer we compare two actual cases that came under the authors' observation. The mixers used were of the same size and make. In one case the stone was shoveled into the charging hopper by four men and the sand and cement were delivered in barrows by four other men; six men took the concrete away in wheelbarrows. The output of the mixer was one batch every 5 minutes, or 120 batches, or 60 cu. yds., in 10 hours. In the other case the sand and the stone were chuted directly into the charging hopper from overhead bins and the mixer discharged into one-batch buckets on cars. The output of the mixer was one batch every 2 minutes, or 300 batches in 10 hours. In the first case the capacity of the mixer was limited by the ability of a gang of workable size to get the raw materials to and the mixed concrete away from the mixer. In the second case the capacity was limited only by the amount of mixing deemed necessary.

While the necessity of rapid charging of a mixer to secure its best output is generally realized it is often forgotten that the rapidity of discharge is also a factor of importance. The size of the conveyor by which the concrete is removed affects the time of discharge. By timing a string of wheelbarrows in line the authors have found that it takes about 7 seconds to fill each barrow; as a rule slight delays will increase this time to 10 seconds. With a load of 1 cu. ft. per barrow it requires 13 barrow loads to take away a ½ cu. yd. batch. This makes the time of discharging a batch 130 seconds, or say 2 minutes. The same mixer discharging into a batch size bucket will discharge in 15 to 20 seconds, saving at least 1½ minutes in discharging each batch.

MIXER EFFICIENCY.—Various attempts have been made to rate the efficiency of concrete mixers. In all cases a percentage basis of comparison has been adopted; arbitrary values are assigned to the several functions of a mixer, such as 40 per cent. for perfect mixing, 10 per cent. for time of mixing and 25 per cent. for control of water, the total being 100 per cent., and each mixer analyzed and given a rating according as it is considered to approach the full value of any function. Such percentage ratings are unscientific and misleading; they present definite figures for what are mere arbitrary determinations. The values assigned to the several functions are purely arbitrary in the first place, and in the second place the decision as to how near those values any mixer approaches are matters of personal judgment.

The most efficient mixer is the one that gives the maximum product of standard quality at the least cost for production.

This rule recognizes the fact that in practical construction different standards of quality are accepted for different kinds of work. No engineer demands, for example, the same quality of mixture for a pavement base that he does for a reinforced concrete girder. If mixer A turns out concrete of a quality suitable for pavement base cheaper than does mixer B, then it is the more efficient mixer for the purpose, even though mixer B will make the superior quality of concrete required for a reinforced girder while mixer A will not. This method of determining efficiency holds accurate for any standard of quality that may be demanded.

5

UNDER WATER CONCRETE AND SUBAQUEOUS GROUTING

Mixed concrete if emptied loose and allowed to sink through water is destroyed; the cement paste is washed away and the sand and stone settle onto the bottom more or less segregated and practically without cementing value. In fact, if concrete is deposited with the utmost care in closed buckets and there is any current to speak of a considerable portion of cement is certain to wash out of the deposited mass. Even in almost still water some of the cement will rise to the surface and appear as a sort of milky scum, commonly called *laitance*. Placing concrete under water, therefore, involves the distinctive task of providing means to prevent the washing action of the water. It is also distinguished from work done in air by the fact that it cannot be compacted by ramming, but the main problem is that of preventing wash during and after placing.

DEPOSITING IN CLOSED BUCKETS.—Special buckets for depositing concrete under water are made by several manufacturers of concrete buckets. These buckets vary in detail but are all similar in having doors to close the concrete away from the water and, generally, in being bottom dumping.

The bucket shown by Fig. 26 was designed by Mr. John F. O'Rourke, and is built by the Cockburn Barrow & Machine Co., of Jersey City, N. J. This bucket was used in depositing the concrete for the City Island Bridge foundations described in Chapter XII and also in a number of other works. It consists of a nearly cubical shell of steel open at top and bottom, and having heavy timbers rivetted around the bottom edges. The open top has two flat flap doors. Two similar doors hinged about midway of the sides close to form a V-shaped hopper bottom inside the shell and serve when open, to close the openings in the sides of the shell. In loading the bucket the bottom doors are drawn inward and upward by the chains and held by a temporary key. The loaded bucket is then lifted by the bail and the key removed, since when suspended the pull on the bail holds the chains taut and the doors closed. As soon as the bucket rests on the bottom the pull of the concrete on the doors slides the bail down and the doors swing downward and back discharging the concrete. The timbers around the bottom edges keep the bucket from sinking into the deposited concrete, and the doors and shell exclude all water from the batch until it is finally in place.

Fig. 26.—O'Rourke Bucket fur Depositing Concrete Under Water.

The subaqueous concrete bucket shown by Figs. 27 and 28 is made by the Cyclopean Iron Works Co., Jersey City, N. J. Fig. 27 shows the bucket suspended full ready for lowering; the cover is closed and latched and the bail is held vertical by the tag line catch *A*. Other points to be noted are the eccentric pivoting of the bail, the latch unlocking lever and roller *B* and *C*, and the stop *D*. In the position shown the bucket is lowered through the water and when at the proper depth just above bottom the tag line is given a sharp pull, uncatching the bail. The body of the bucket turns bottom side up, revolving on the bail pivots, and just as the revolution is completed the bail engages the roller *C* on the latch unlocking lever and swings the lever enough to unlatch the top and allow it to swing down as shown by Fig. 28 and release the concrete. The stop *D* keeps the body of the bucket from swinging beyond the vertical in dumping.

Fig. 27.—Cyclopean Bucket for Depositing Concrete Under Water (Closed Position).

Fig. 28.—Cyclopean Bucket for Depositing Concrete Under Water (Open Position).

Figures 29 and 30 show the subaqueous concrete bucket made by the G. L. Stuebner Iron Works, Long Island City, N. Y., essentially the same bucket, omitting the cover and with a peaked bail, is used for work in air. For subaqueous work the safety hooks *A* are lifted from the angles *B* and wired to the bail in the position shown by the dotted lines, and a tag line is attached to the handle bar *C*. The bucket being filled and the cover placed is lowered through the water to the bottom and then discharged by a pull on the tag line.

DEPOSITING IN BAGS.—Two methods of depositing concrete in bags are available to the engineer; one method is to employ a bag of heavy tight woven material, from which the concrete is emptied at the bottom, the bag serving like the buckets previously described simply as means of conveyance, and the other method is to use bags of paper or loose woven gunnysack which are left in the work, the idea being that the paper will soften or the cement will ooze out through the openings in the cloth sufficiently to bond the separate bagfuls into a practically solid mass.

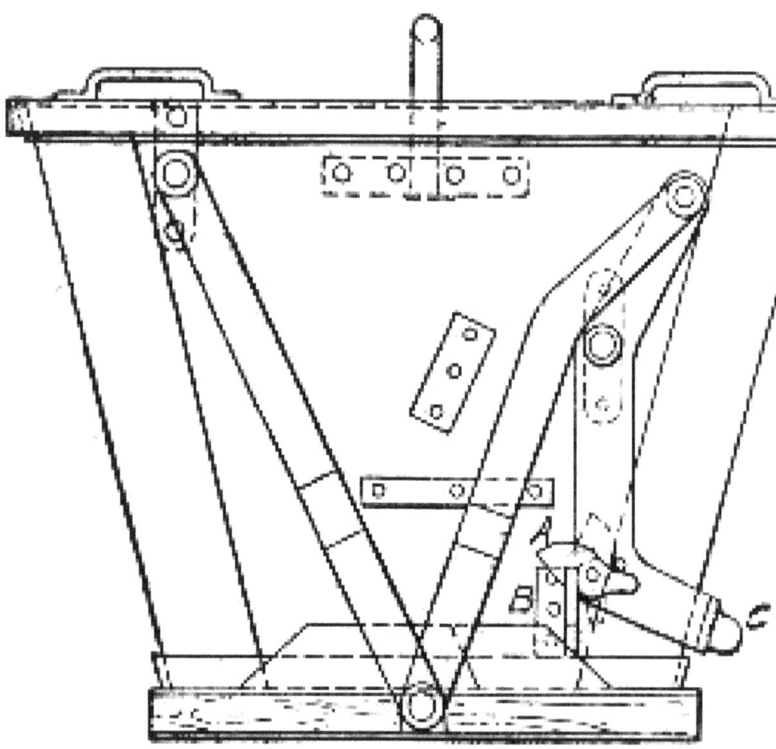

Fig. 29.—Stuebner Bucket for Depositing Concrete Under Water (Closed Position).

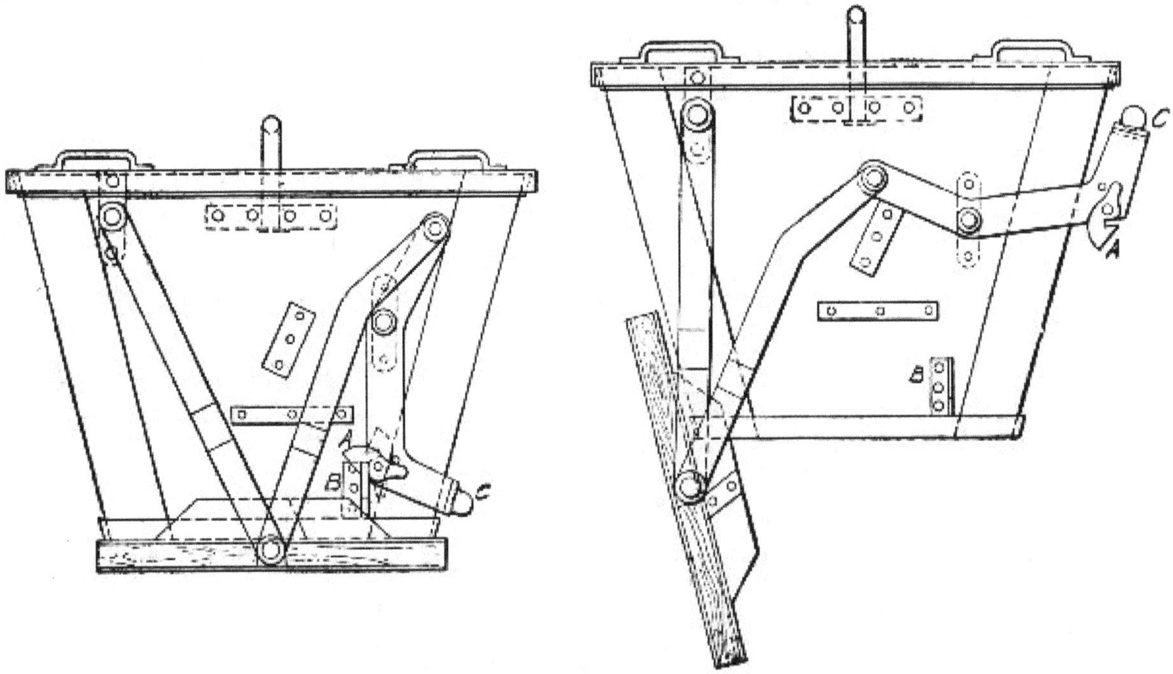

Fig. 30.—Stuebner Bucket for Depositing Concrete Under Water (Open Position).

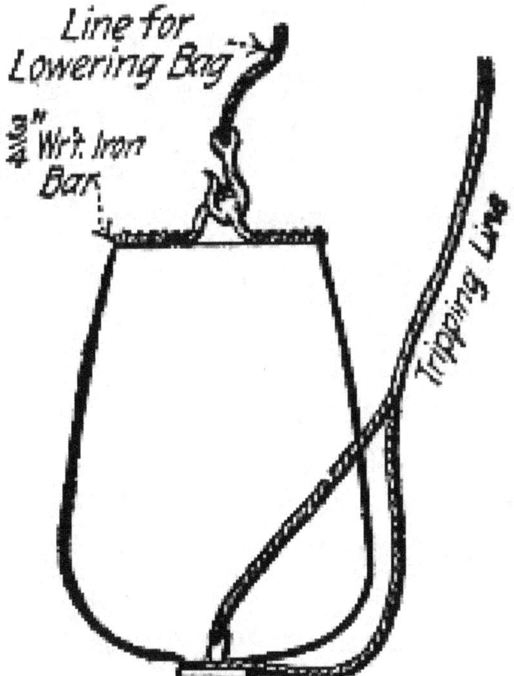

Fig. 31.—Bag for Depositing Concrete Under Water.

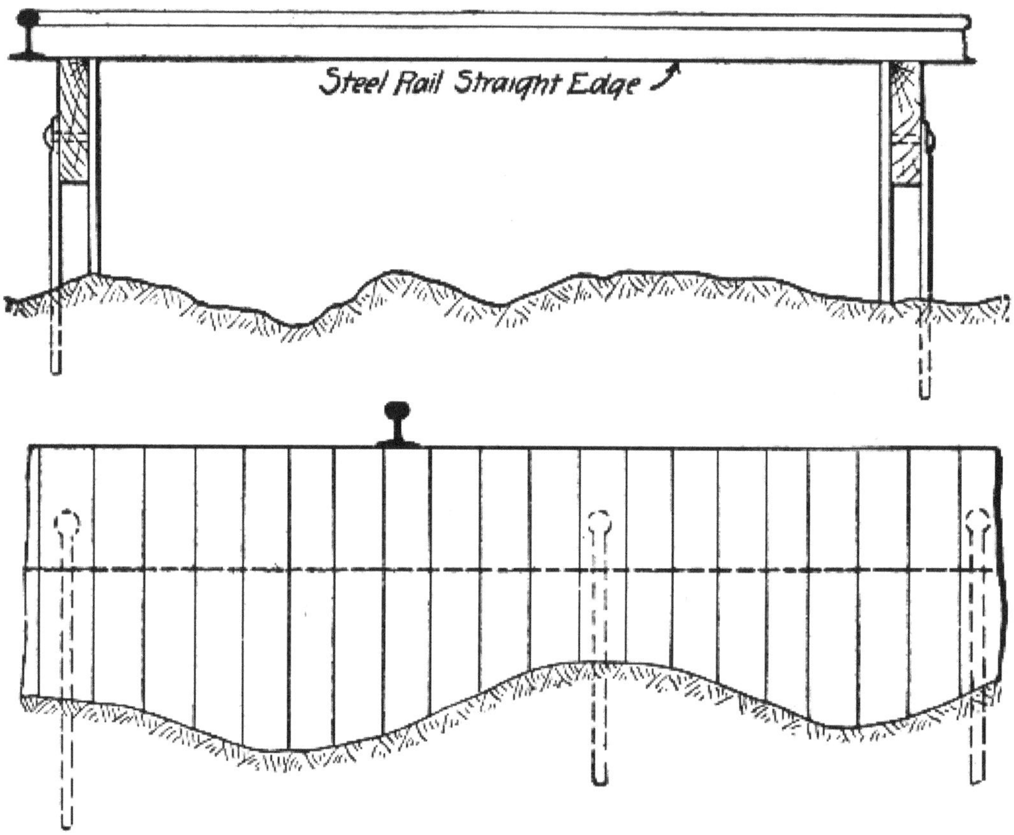

Fig. 32.—Form for Molding Footing for Block Concrete Breakwater.

The bag shown by Fig. 31 was used to deposit concrete for leveling up a rough rock bottom and so provide a footing for a concrete block pier constructed in 1902 at Peterhead, N. B., by Mr. William Shield, M. Inst. C. E. Careful longitudinal profiles were taken of the rock bottom one at each edge of the footing. Side forms were then made in 20-ft. sections as shown by Fig. 32; the lagging boards being cut to fit the determined profile and the top of the longitudinal piece being flush with the top of the proposed footing. The concrete was filled in between the side forms and leveled off by the T-rail straight-edge. In placing the side forms the longitudinal pieces were placed by divers who were given the proper elevations by level rods having 10 to 15-ft. extension pieces to raise the targets above the water surface. When leveled the side pieces were anchor-bolted as shown to the rock, the anchor-bolts being wedged into the holes to permit future removal. The concrete was then lowered in the bag shown by Fig. 31, the divers assisting in guiding the bag to position. The mouth of the bag being tied by one turn of a line having loops through which a wooden key is slipped to hold the line tight, a sharp tug on the tripping rope loosens the key and empties the bag. The bags used on this work had a capacity of 2¼ cu. ft. To permit the removal of the side forms after the concrete had hardened, a strip of jute sacking was spread against the lagging boards with a flap extending 15 to 18 ins. under the concrete. The forms were removed by divers who loosened the anchor bolt wedges.

In placing small amounts of concrete for bridge foundations in Nova Scotia, bags, made of rough brown paper were used to hold the concrete. Each bag held about 1 cu. ft. The bags were made up quickly and dropped into the water one after the other so that the following one was deposited before the cement escaped from the former one. The paper was immediately destroyed by submersion and concrete remained. The bags cost $1.35 per hundred or 35 cts. per cu. yd. of concrete. Concrete was thus deposited in 18 ft. of water without a diver.

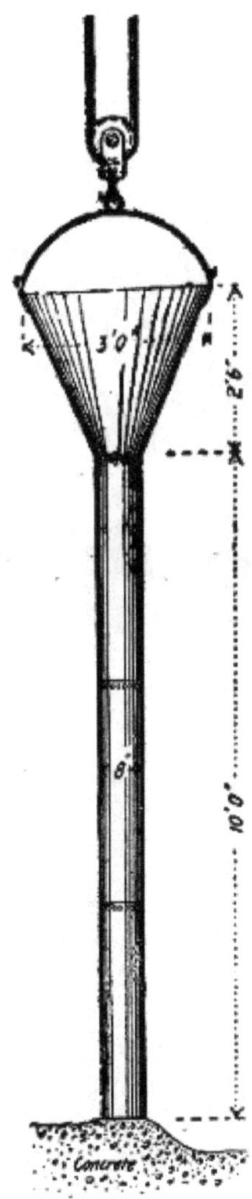

Fig. 33.—Steel Tremie for Depositing Concrete Under Water.

DEPOSITING THROUGH A TREMIE.—A tremie consists of a tube of wood or, better, of sheet metal, which reaches from above the surface to the bottom of the water; it is operated by filling the tube with concrete and keeping it full by successive additions while allowing the concrete to flow out gradually at the bottom by raising the tube slightly to provide the necessary opening. A good example of a sheet steel tremie is shown by Fig. 33. This tremie was used by Mr. Wm. H. Ward in constructing the Harvard Bridge foundations and numerous other subaqueous structures of concrete.

In these works the tube was suspended from a derrick. Wheelbarrows filled the tube and hopper with concrete and kept them full; the derrick raised the tube a few inches and swung it gently so as to move it slowly over the area to be filled. Care being taken to keep the tube at one height, the concrete was readily deposited in even layers. Concrete thus deposited in 18 ft. of water was found to be level and solid on pumping the pit dry.

Fig. 34.—Tremie and Traveler Used at Charlestown, Mass., Bridge.

Another method of handling a tremie was employed in constructing the foundations for the Charlestown Bridge at Boston, Mass. Foundation piles were driven and sawed off under water. A frame was built above water and supported by a curbing attached to certain piles in the outer rows of the foundation reserved for this purpose. In this frame the vertical members were Wakefield sheet-piling plank, spaced 6 to 10 ft. apart, and connected by three lines of double waling bolted to the verticals at three different heights. This frame was lowered to the bottom so as to enclose the bearing piles. The posts or verticals were then driven, one by one, into the bottom, the frame being flexible enough to permit this. The spaces between the posts or verticals were then filled by sheet-piling and the frame was bolted to the curbing piles. This curbing afterward supported the traveler used in laying the concrete. Thus a coffer dam was formed to receive the concrete as shown in Fig. 34. The 1-2-5 concrete was deposited up to within 5½ ft. of the mean low water level, the last foot being laid after water was pumped out. The tremie used to deposit the concrete was a tube 14 ins. in diameter at the bottom and 11 ins. at the neck, with a hopper at the top. It was made in removable sections, with outside flanges, and was suspended by a differential hoist from a truck moving laterally on a traveler, Fig. 34. The foot of the chute rested on the bottom until filled with concrete; then the chute was slowly raised and the concrete allowed to run but into a conical heap, more concrete being dumped into the hopper. As the truck moved across the traveler a ridge of concrete was made; then the traveler was moved forward and another parallel ridge was made. The best results were obtained when the layers were 2½ ft. thick, but layers up to 6 ft. thick were laid. If the layer was too thick, or uneven, or if the chute was moved or raised too quickly, the charge in the tube was "lost." This was objectionable because the charging of the chute anew resulted in "washing" the cement more or less out of the concrete until the chute was again filled. To reduce this objection the contractor was directed to dump some neat cement into the tube before filling with concrete. A canvass piston was devised which could be pushed ahead of the concrete when filling the chute. It consisted of two truncated cones of canvass, one flaring downward to force the water ahead, and the other flaring upward to hold the concrete. The canvass was stiffened and held against the sides of the chute by longitudinal ribs of spring steel wire; the waist was filled by a thick block of wood to which all the springs were attached; and to this block were connected additional steel guides to prevent overturning and a rope to regulate the descent. Very little water forced its way past this piston and it was a success, but as the cost was considerable and a piston was lost each time, its use was abandoned as the evil to be avoided did not justify the outlay.

The chute worked best when the concrete was mixed not quite wet enough to be plastic. If mixed too wet the charge was liable to be "lost," and if dry it would choke the chute. An excess of gravel permitted water to ascend in the tube; and an excess of sand tended to check the flow of concrete.

In constructing the piers for a masonry arch bridge in France in 1888 much the same method was followed, except that a wooden tremie 16 ins. square made in detachable sections was used. This tremie had a hopper top and was also provided with a removable cap or cover for the bottom end, the latter device being intended to keep the water out of the tube and prevent "washing" the first charge of concrete. The piers were constructed by first driving piles and sawing them off several feet above the bottom but below water level, and then filling them nearly to their tops with broken stone. An open box caisson was then sunk onto the stone and embracing the pile tops and then filled around the outside with more broken stone. The caisson was then filled with concrete through the tremie which was handled by a traveling crane. The crane was mounted and traveled transversely of the pier on a platform which in turn moved along tracks laid lengthwise of the caisson. The tube was gradually filled with concrete and lowered, the detachable bottom of the tube was then removed, allowing the concrete to run out. The tube was first moved across the caisson and then downstream and back across the caisson, and this operation repeated until a 16-in. layer was completed. The tube was then raised 16 ins. and the operations repeated to form another layer. There was almost no *laitance*. From 90 to 100 cu. yds. were deposited daily.

Still another example of tremie work is furnished by the task of depositing a large mass of concrete under water in the construction of the Nussdorf Lock at Vienna. This lock has a total width of 92 ft. over all, and is 49.2 ft. clear inside. The excavation, which was carried to a depth of 26.24 ft. below water level, was made full width, between sheet piling, and the bottom was filled in with rammed sand and gravel, forming a kind of invert with its upper surface horizontal in the middle and sloping upwards a trifle at both sides. A mass of concrete having a total thickness of 13.12 ft. was built on this foundation in the center where the upper surfaces were 13.12 ft. below the water level. Concrete walls were carried up at the sides of the lock to a height of 3.28 ft.; these walls were 8.2 ft. thick. The methods used in placing the concrete were as follows: Three longitudinal rows of piles were driven on each side of the axis of the lock, these piles supporting a 6-rail track about 7 ft. above the water level. Three carriages spanning the full width of the lock transversely moved on this track. Each carriage had three trolleys, one in each of the main panels of the transverse pile bends. These trolleys each carried a vertical telescopic tube, by means of which the concrete was deposited at the bottom of the lock. These tubes or chutes were of different lengths in the three carriages; the first ones deposited the concrete up to a level of 23 ft. below the surface; the next set deposited the concrete between that level and 19.7 ft., and the last set completed the subaqueous work up to the final height of 16.4 ft. below the surface. The tops of the tubes were level with a transverse track extending the full length of the carriage. The ends of these tracks just cleared the outside rows of piles, which, on one side of the lock, supported a distribution track parallel to the axis of the lock. Dump cars running on this distribution track delivered the concrete to smaller dump cars on the carriage tracks, and in turn these smaller cars dumped into either of these chutes on each carriage. The carriages were moved from end to end of the lock, the whole area of the lock coming under the nine chutes, inasmuch as each chute moved one-third the length of the carriage. The concrete was deposited in three horizontal layers 3.28 ft. thick, the layers being built in comparatively narrow banks, so that the different layers would key together and form a corrugated mass. The chutes were shortened as the concrete was deposited, three layers being placed successively. The main body of the bottom and the side walls were built by this method, and then the water was pumped out and a 2.3 ft. layer of concrete rammed over the bottom and completed with a finished surface 9 ft. thick.

GROUTING SUBMERGED STONE.—Masses of gravel, broken or rubble stone deposited under water may be cemented into virtually a solid concrete by charging the interstices with grout forced through pipes from the surface. Mr. H. F. White gives the following records of grouting submerged gravel:

In experiment No. 1 a reservoir 10 ft. square was filled to a depth of 18 ins. with clean gravel ballast (1½ to 2-in. size) submerged in water. A 2-in. gas pipe rested on the gravel and was surmounted with a funnel. A 1:1 Portland grout was poured in. After 21 days set the water was drawn off, and it was found that the grout had permeated the ballast for a space of 8 ft. square at the bottom and 6 ft. square at the top, leaving a small pile of pure cement mortar 6 ins. high about the base of the pipe; 16 cu. ft. of cement and 16 cu. ft. of sand concreted 100 cu. yds. of ballast. In experiment No. 2, under the same conditions, a grout made of 1 part lime, 1 part surki (puzzulana or trass) and 1 part sand, was found to have spread over the entire bottom, 10 ft. square, rising 5 ins. on the sides, and making the concreted mass about 3½ ft. square at the top; 25 cu. ft. of the dry materials concreted 100 cu. ft. of ballast. In experiment No. 3 the ballast was 2½ ft. deep. A grout (using 8 cu. ft. of each ingredient) made as in experiment No. 2 covered the bottom, rose 14 ins. on the sides and made a top surface 4½ ft. square; 32 cu. ft. of the dry materials grouted 100 cu. ft. of ballast. In experiment No. 4 the ballast was of bats and pieces 3 or 4 ins. in size laid 7 ft. deep. A grout made as in experiment No. 2 (using 88 cu. ft. of each ingredient) concreted the whole mass to a depth of 6 ft. up the sides, and 2½ ft. square at the pipe on the surface of the ballast. Mr. White says that a grout containing more than 1 part of sand to 1 of Portland cement will not run freely through a 2-in. pipe, as the sand settles out and chokes the pipe. Even with 1:1 grout it must be constantly stirred and a steady flow into the pipe maintained. The lime-trass grout does not give the same trouble.

Mr. W. R. Knipple describes the work of grouting rubble stone and gravel for the base of the Hermitage Breakwater. This breakwater is 525 ft. long, 50 ft. wide at base and 42 ft. wide at top, and 68 ft. high, was built on the island of Jersey. Where earth (from 0 to 8½ ft. deep) overlaid the granite rock, it was dredged and the trench filled in with rubble stones and gravel until a level foundation was secured. Cement grout was then forced into this filling through pipe placed 8 to 10 ft. apart. The grouting was done in sections 12½ ft. long, from 7 to 10 days being taken to complete each. Upon this foundation concrete blocks, 4×4×9 to 12 ft., were laid in courses inclined at an angle of 68°. The first four courses were laid by divers, the blocks being stacked dry two courses high at a time. The joints below water were calked by divers and above water by masons, and a section was then grouted. When two courses had been laid and grouted, two more courses were laid and grouted in turn, and so on. In places, grouting was done in 50 ft. of water. The grout should be a thick paste; a 30-ft. column of grout will balance a 60-ft. column of water.

6

RUBBLE AND ASPHALTIC CONCRETE

Two kinds of concrete which vary in composition and character from the common standard mixtures of cement, sand and broken aggregate are extensively employed in engineering construction. These are rubble concrete and asphaltic concrete.

RUBBLE CONCRETE.—In constructing massive walls and slabs a reduction in cost may often (not always) be obtained by introducing large stones into the concrete. Concrete of this character is called rubble concrete, and the percentage of rubble stone contained varies from a few per cent. to, in some cases, over half of the volume. The saving effected comes partly from the reduction in the cement required per cubic yard of concrete and partly from the saving in crushing.

The saving in cement may be readily figured if the composition of the concrete and the volume of the added rubble stones be known. A 1-2½-5 concrete requires according to Table X in Chapter II 1.13 bbls. of cement per cubic yard. Assuming a barrel of cement to make 3.65 cu. ft. of paste, we have 3.65 × 1.13 = 4.12 cu. ft. of cement paste per cubic yard of 1-2½-5 concrete. This means that about 15 per cent. of the volume of the concrete structure is cement. If rubble stone be introduced to 50 per cent. of the volume, then the structure has about 7½ per cent. of its volume of cement. It is of interest to note in this connection that rubble masonry composed of 65 per cent. stone and 35 per cent. of 1-2½ mortar would have some 11½ per cent. of its volume made up of cement.

The saving in crushing is not so simple a determination. Generally speaking, the fact that a considerable volume of the concrete is composed of what, we will call uncrushed stone, means a saving in the stone constituent of one structure amounting to what it would have cost to break up and screen this volume of uncrushed stone, but there are exceptions. For example, the anchorages of the Manhattan Bridge over the East River at New York city were specified to be of rubble concrete, doubtless because the designer believed rubble concrete to be cheaper than plain concrete. In this case an economic mistake was made, for all the rubble stone used had to be quarried up the Hudson River, loaded onto and shipped by barges to the site and then unloaded and handled to the work using derricks. Now this repeated handling of large, irregular rubble stones is expensive. Crushed stone as we have shown in Chapter IV can be unloaded from boats at a very low cost by means of clam shells. It can be transported on a belt conveyor, elevated by bucket conveyer, mixed with sand and cement and delivered to the work all with very little manual labor when the installation of a very efficient plant is justified by the magnitude of the job. Large rubble stones cannot be handled so cheaply or with so great rapidity as crushed stone; the work may be so expensive, due to repeated handlings, as to offset the cost of crushing as well as the extra cost of cement in plain concrete. On the other hand, the cost of quarrying rock suitable for rubble concrete is no greater than the cost of quarrying it for crushing—it is generally less because the stone does not have to be broken so small—so that when the cost of getting the quarried rock to the crusher and the crushed stone into the concrete comes about the same as getting the quarried stone into the structure it is absurd practice to require crushing. To go back then to our first thought, the question whether or not saving results from the use of rubble concrete, is a separate problem in engineering economics for each structure.

In planning rubble concrete work the form of the rubble stones as they come from the quarry deserves consideration. Stones that have flat beds like many sandstones and limestones can be laid upon layers of dry concrete and have the vertical interstices filled with dry concrete by tamping. It requires a sloppy concrete to thoroughly embed stones which break out irregularly. In the following examples of rubble concrete work the reader will find structures varying widely enough in character and in the percentages of rubble used to cover most ordinary conditions of such work.

Where the rubble stones are very large it is now customary to use the term "cyclopean masonry" instead of rubble concrete. Many engineers who have not studied the economics of the subject believe that the use of massive blocks of stone bedded in concrete necessarily gives the cheapest form of masonry. We have already indicated conditions where ordinary concrete is cheaper than rubble concrete. We may add that if the quarry yields a rock that breaks up naturally into small sized blocks, it is the height of economic folly to specify large sized cyclopean blocks. Nevertheless this blunder has been frequently made in the recent past.

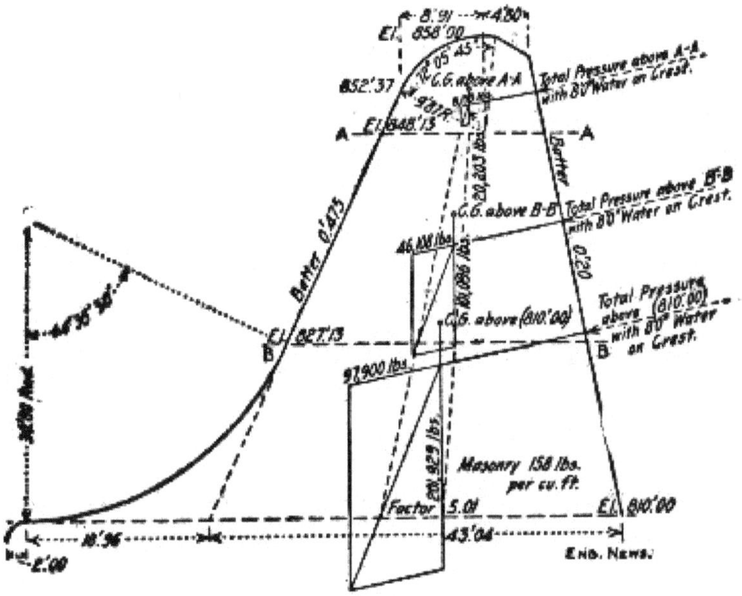

Fig. 35.—Diagram Cross-Section of Rubble Concrete Dam, Chattahoochee River.

Chattahoochee River Dam.—The roll-way portion, 680 ft. long, of the dam for the Atlanta Water & Electric Power Co., shown in section by Fig. 35, was built of a hearting of rubble concrete with a fine concrete facing and a rubble rear wall. The facing, 12 ins. thick of 1-2-4 concrete, gave a smooth surface for the top and face of the dam, while the rubble rear wall enabled back forms to be dispensed with and, it was considered, made a more impervious masonry. The concrete matrix for the core was a 1-2-5 stone mixture made very wet. The rubble stones, some as large as 4 cu. yds., were bedded in the concrete by dropping them a few yards from a derrick and "working" them with bars; a well formed stone was readily settled 6 ins. into a 10-in. bed of concrete. The volume of rubble was from 33 to 45 per cent. of the total volume of the masonry. The 1-2-4 concrete facing was brought up together with the rubble core, using face forms and templates to get the proper profile. The work was done by contract and the average was 5,500 cu. yds. of concrete placed per month.

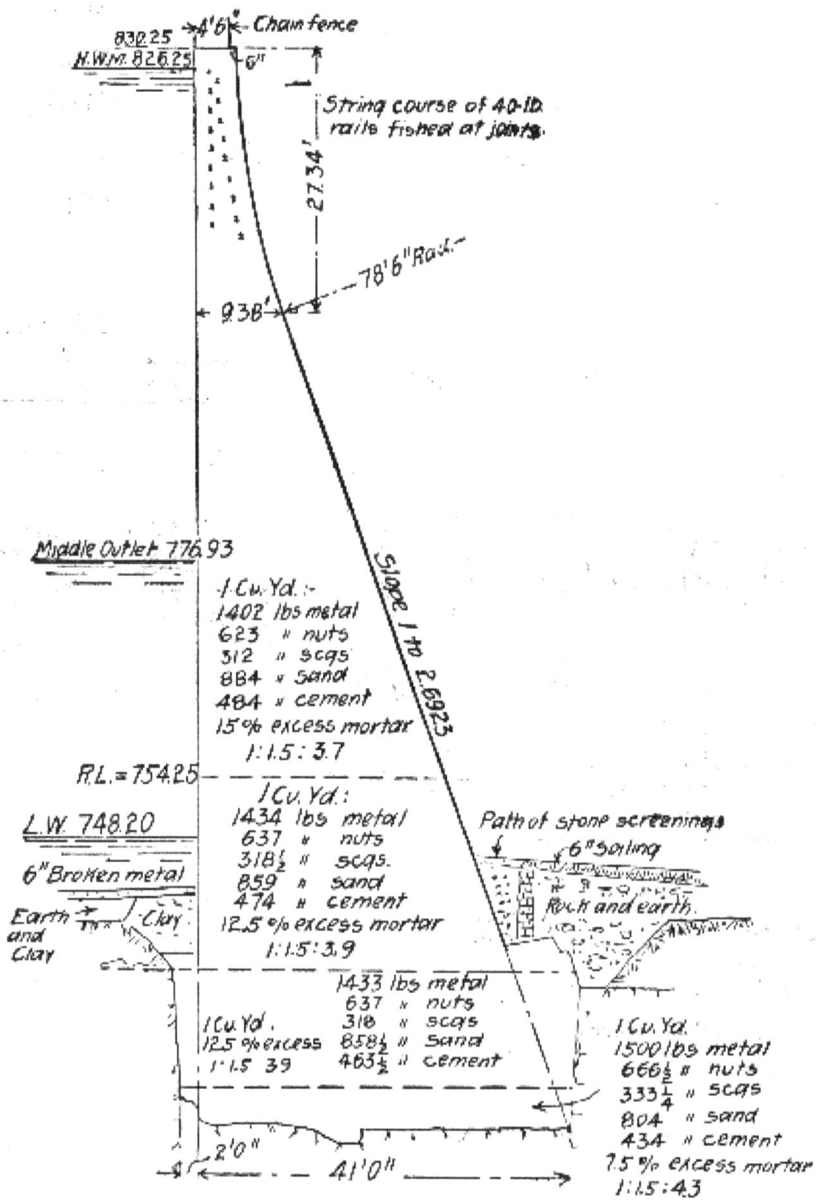

Fig. 36.—Cross-Section of Barossa Dam of Rubble Concrete.

Barossa Dam, South Australia.—The Barossa Dam for the water-works for Gawler, South Australia, is an arch with a radius of 200 ft., and an arc length on top of 422 ft.; its height above the bed of the stream is 95 ft. Figure 36 is a cross-section of the dam at the center. The dam contains 17,975 cu. yds. of rubble concrete in the proportions of 2,215 cu. yds. of rubble stone to 15,760 cu. yds. of concrete; thus about 12.3 per cent. of the dam was of rubble. The concrete was mixed by weight of 1 part cement, 1½ parts sand, and a varying proportion of aggregate composed of 4½ parts 1¼ to 2-in. stone, 2 parts ½ to 1¼-in. stone and 1 part ⅛ to ½-in. stone or screenings. The sand was one-half river sand and one-half crusher sand. The following shows the amounts by weight of the several materials for each of the several classes of concrete per cubic yard:

Class.	Excess Mortar.	Stone 1¼-2.	Stone ½-1¼.	Stone ⅛-½.	Sand.	Cement.
A	7.5%	1,500	661½	333¼	804	434
B	12.5	1,433⅓	637	318	858½	463
C	12.5	1,434	637	318½	859	474
D	15	1,402	623	312	884	484

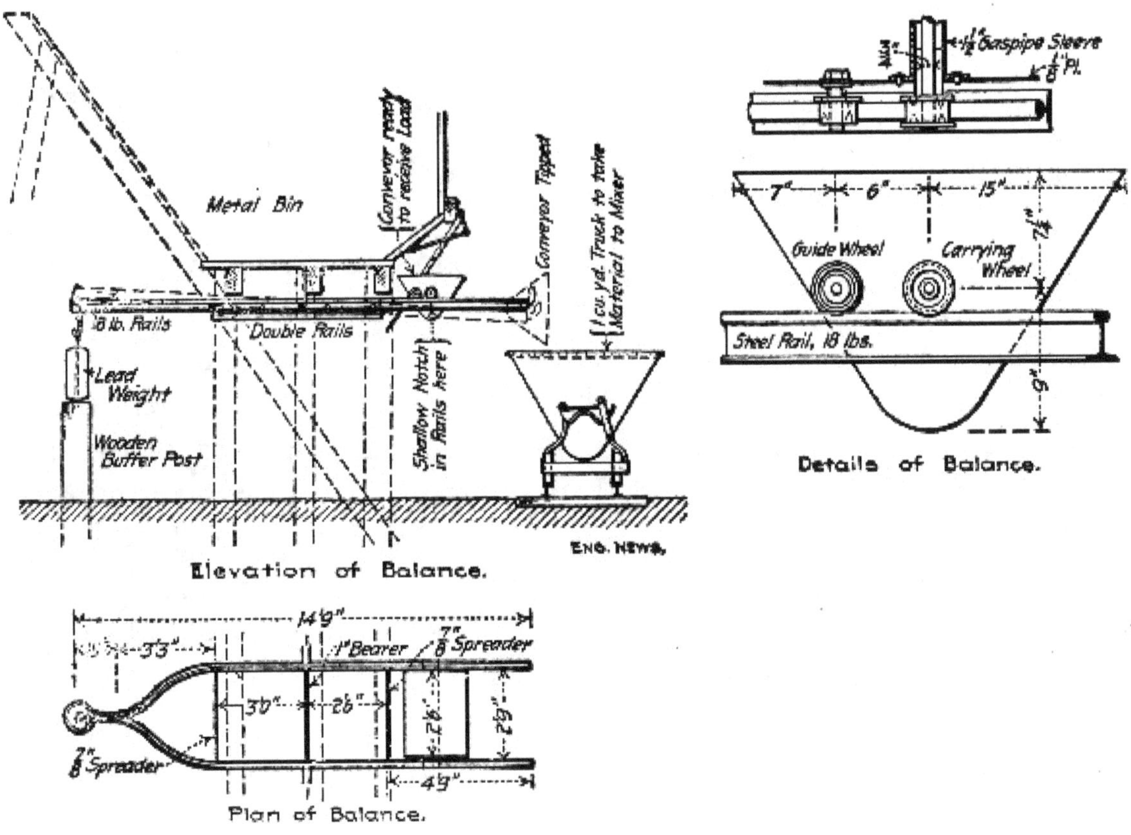

Fig. 37.—Apparatus Used for Weighing Concrete Materials at Barossa Dam.

The average composition of the concrete was 1-1½-3½. Its cost per cubic yard in place including rubble was 38s 9d per cu. yd. or about $9.30. In proportioning the mixture on the work use was made of the device shown by Fig. 37 to weigh the aggregate. The measuring car is pushed back under the stone hopper chute until the wheels drop into shallow notches in the balanced track rails; stone is then admitted until the lead weight begins to rise, when the car is pushed forward and dumps automatically as indicated.

Other Rubble Concrete Dams.—Rubble concrete containing from 55 to 60 per cent. rubble was used in constructing the Boonton Dam at Boonton, N. J. The stones used measured from 1 to 2½ cu. yds. each; the concrete was made so wet that when the stones were dropped into it, it flowed into every crevice. The materials were all delivered on cars, from which they were delivered to the dam by derricks provided with bull-wheels. On the dam there were 4 laborers and 1 mason to each derrick, and this gang dumped the concrete and joggled the rubble stones into it. Records of 125 cu. yds. per 10 hours, with one derrick, were made. With 35 derricks, 20 of which were laying masonry and 15 either passing materials or being moved, as much as 21,000 cu. yds. of masonry were laid in one month. The amount of cement per cubic yard of masonry is variously stated to have been 0.6 to 0.75 bbl. The stone was granite.

The Spier Falls Dam on the upper Hudson River was built of rubble concrete containing about 33 per cent. rubble stone. The concrete was a 1-2½-5 mixture, and the engineer states that about 1 bbl. of cement was used per cubic yard of rubble concrete. This high percentage of cement may be accounted for by the fact that there was a considerable amount of rubble masonry in cement mortar included in the total. The stones and concrete were delivered along the dam by cableways and stiff-leg derricks set on the downstream sloping face of the dam delivered them from the cableways into place. There were two laborers to each mason employed in placing the materials, wages being 15 and 35 cts. per hour, respectively. The labor cost of placing the materials was 60 cts. per cubic yard of masonry. The stone was granite.

Granite rubble laid in layers on beds of concrete and filled between with concrete was used in constructing the Hemet Dam in California. The concrete was a 1-3-6 mixture, and was thoroughly tamped under and between the stones. For face work the stones were roughly scabbled to shape and laid in mortar. The stone was taken from the quarry 400 ft. away and delivered directly on the dam by cableways; here two derricks handled the stone into place, the dam being only 246 ft. arc length on top, though it was 122½ ft. high. The cableways would take a 10-ton load; stones could be taken from the quarry, hoisted 150 ft. and delivered to the work in 40 to 60 seconds. Common labor at $1.75 per day was used for all masonry except facing, where masons at $3.50 were employed. Cement cost delivered $5 per barrel, of which from $1 to $1.50 per barrel was the cost of hauling 23 miles by team over roads having 18 per cent. grades in places. Sand was taken from the stream bed and delivered to the work by bucket conveyor. "Under favorable conditions some of the masonry was put in for as low as $4 per cu. yd." There were 31,100 cu. yds. of masonry in the dam, which required 20,000 bbls. of cement, or 0.64 bbl. per cubic yard.

The following novel method of making rubble concrete was employed in enlarging two old dams and in constructing two new dams for a small water-works. The available time was short, the amount of work was too small and too scattered to justify the installation of a stone crusher, and suitable gravel was not at hand. Sufficient small boulders in old walls, and borrow pits and on surface of fields were available, and were used with thin Portland cement mortar. One part of Alpha or Lehigh cement and three parts sand were mixed dry at first and then wet with just enough water to make the resulting mortar flow by gravity. This mortar was shoveled into the forms continuously by one set of men while other men were throwing into the mortar in the forms the boulders which were cleaned and broken so as not to be more than 7 ins. long. In general the performance was continuous. Three mortar beds were placed parallel with, and against, one side of the forms, with spaces of about 4 ft. between the ends of the beds. The boulders were dumped on the opposite side of the forms. Two men shoveled in all the mortar and did nothing else. While they were emptying one bed the mortar was being mixed in the preceding bed by two other men and the materials placed in the third bed by still others. Another gang was continually throwing in the boulders and small stones and still another was breaking stone. One man should keep the mortar well stirred while the bed is being emptied. About 20 men were necessary to do all parts of the work. The forms were of 2-in. planed plank tongued and grooved. Especial pains were taken to make the forms tight, and all leaks that appeared were quickly stopped with dry cement. Some pains were taken to prevent a flat side of large stones from coming in direct contact with the forms, but round boulders and small stones needed no care to prevent their showing in the finished work.

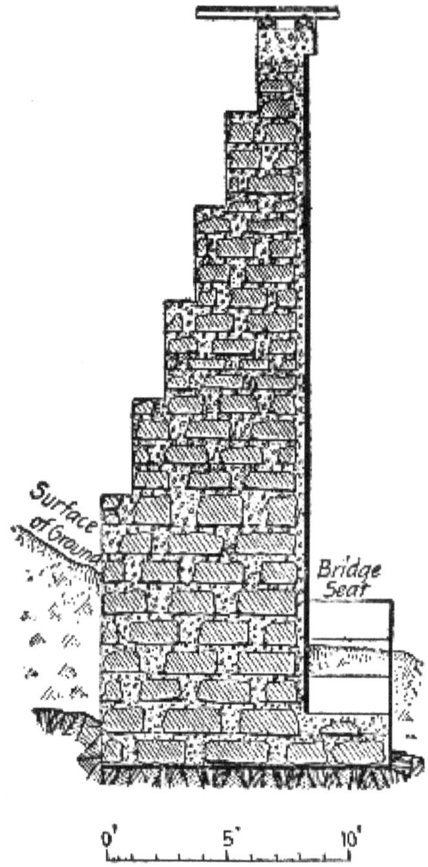

Fig. 38.—Bridge Abutment of Rubble Concrete.

In conclusion it is interesting to note, perhaps, the earliest use of rubble concrete for dam construction in this country in constructing the Boyd's Corner Dam on the Croton River near New York. This dam was begun in 1867 and for a time rubble concrete was used, but was finally discontinued, due to the impression that it might not be watertight. The specifications called for dry concrete to be thoroughly rammed in between the rubble stones, and to give room for this ramming the contractor was not permitted to lay any two stones closer together than 12 ins. As a result not more than 33 per cent. of the concrete was rubble.

Abutment for Railway Bridge.—Figure 38 shows a bridge abutment built of rubble concrete at a cost of about $4.50 per cu. yd. The concrete was a 1-2½-4½ mixture laid in 4-in. layers. On each layer were laid large rubble stones bedded flat and spaced to give 6-in. vertical joints; the vertical joints were filled with concrete by ramming and then another layer of concrete placed and so on. A force of 28 men and a foreman averaged 40 cu. yds. of rubble concrete per day. The following is the itemized cost per cubic yard, not including forms, for 278 cu. yds:

Item.	Per Cu. Yd.
0.82 bbls. cement, at $2.60	$2.14
0.22 cu. yd. sand, at $1.00	0.22
0.52 cu. yd. broken stone, at $0.94	0.49
0.38 cu. yd. rubble stone, at $0.63	0.24
Water	0.07
Labor, at 15 cts. per hour	1.19
Foreman	0.09
Total	$4.44

Some English Data on Rubble Concrete.—Railway work, under Mr. John Strain, in Scotland and Spain, involved the building of abutments, piers and arches of rubble concrete. The concrete was made of 1 part cement to 5 parts of ballast, the ballast consisting of broken stone or slag and sand mixed in proportions determined by experiment. The materials were mixed by turning with shovels 4 times dry, then 4 times more during the addition of water through a rose nozzle. A bed of concrete 6 ins. thick was first laid, and on this a layer of rubble stones, no two stones being nearer together than 3 ins., nor nearer the forms than 3 ins. The stones were rammed and probed around with a trowel to leave no spaces. Over each layer of rubble, concrete was spread to a depth of 6 ins. The forms or molds for piers for a viaduct were simply large open boxes, the four sides of which could be taken apart. The depth of the boxes was uniform, and they were numbered from the top down, so, that, knowing the height of a given pier, the proper box for the base could be selected. As each box was filled, the next one smaller in size was swung into place with a derrick. The following bridge piers for the Tharsis & Calanas Railway were built:

Name.	Length of Bridge. Ft.	Height of Piers. Ft.	No. of Spans.	Cu. Yds. in Piers.	Weeks to Build.
Tamujoso River	435	28	12	1,737	14½

Oraque	423	31	11	1,590	15
Cascabelero	480	30 to 80	10	2,680	21
No. 16	294	28 to 50	7	1,046	16½
Tiesa	165	16 to 23	8	420	4

It is stated that the construction of some of these piers in ordinary masonry would have taken four times as long. The rock available for rubble did not yield large blocks, consequently the percentage of pure concrete in the piers was large, averaging 70 per cent. In one case, where the stones were smaller than usual, the percentage of concrete was 76½ per cent. In other work the percentage has been as low as 55 per cent., and in still other work where a rubble face work was used the percentage of concrete has been 40 per cent.

In these piers the average quantities of materials per cubic yard of rubble concrete were:

448 lbs. (0.178 cu. yd.) cement.
0.36 cu. yd. sand.
0.68 cu. yd. broken stone (measured loose in piles).
0.30 cu. yd. rubble (measured solid).

Several railway bridge piers and abutments in Scotland are cited. In one of these, large rubble stones of irregular size and weighing 2 tons each were set inside the forms, 3 ins. away from the plank and 3 ins. from one another. The gang to each derrick was: 1 derrick man and 1 boy, 1 mason and 10 laborers, and about one-quarter of the time of 1 carpenter and his helper raising the forms. For bridges of 400 cu. yds., the progress was 12 to 15 cu. yds. a day. The forms were left in place 10 days.

To chip off a few inches from the face of a concrete abutment that was too far out, required the work of 1 quarryman 5 days per cu. yd. of solid concrete chipped off.

Concrete was used for a skew arch over the River Dochart, on the Killin Railway, Scotland. There were 5 arches, each of 30 ft. span on the square or 42 ft. on the skew, the skew being 45°. The piers were of rubble concrete. The concrete in the arch was wheeled 300 ft. on a trestle, and dumped onto the centers. It was rammed in 6-in. layers, which were laid corresponding to the courses of arch stones. As the layers approached the crown of the arch, some difficulty was experienced in keeping the surfaces perpendicular. Each arch was completed in a day.

In a paper by John W. Steven, in Proc. Inst. C. E., the following is given:

	Concrete Per Cu. Yd.	Rubble Concrete Per Cu. Yd.	Per Cent. of Rubble in Rubble Concrete.
Ardrossan Harbor	$6.00	$5.00	20.0
Irvine Branch	7.00	3.68	63.6
Calanas & Tharsis Ry	7.08	3.43	30.3

Mr. Martin Murphy describes some bridge foundations in Nova Scotia. Rubble concrete was used in some of the piers. The rubble concrete consisted of 1 part cement, 2 parts sand, 1 part clean gravel, and 5 parts of large stones weighing 20 lbs. each and upwards. The sand, cement and gravel were turned three times dry and three times wet, and put into the forms. The rubble stones were bedded in the concrete by hand, being set on end, 2 or 3 ins. apart. No rubble stones were placed within 6 ins. of the forms, thus leaving a face of plain concrete; and the rubble stones were not carried higher than 18 ins. below the top of the pier. One cubic yard of this rubble concrete required 0.8 to 0.9 bbl. of cement.

ASPHALT CONCRETE.—Asphalt or tar concrete in which steam cinders or broken stone or gravel and sand are mixed with asphaltum or tar instead of cement paste are used to some extent in lining reservoirs, constructing mill floors, etc. Such mixtures differ in degree only from the mixtures used for asphalt street paving, for discussion of which the various books on paving and asphalts should be consulted. The two examples of asphalt concrete work given here are fairly representative of the mixtures and methods employed for concrete work as distinguished from asphalt work.

Slope Paving for Earth Dam.—Mr. Robert B. Stanton describes a small log dam faced upstream with earth, upon which was laid an asphalt concrete lining to make it water tight. The stone was broken to 2-in. pieces, all the fines being left in and sufficient fine material added to fill the voids. The stone was heated and mixed in pans or kettles from a street paving outfit;

and the asphaltum paste, composed of 4 parts California refined asphaltum and 1 part crude petroleum, was boiled in another kettle. The boiling hot paste was poured with ladles over the hot stone, and the whole mixed over the fire with shovels and hoes. The asphalt concrete was taken away in hot iron wheelbarrows, placed in a 4-in. layer rammed and ironed with hot irons. The concrete was laid in strips 4 to 6 ft. wide, the edges being coated with hot paste. After the whole reservoir was lined, it was painted with the asphalt paste, boiled much longer, until when cold it was hard and stone was broken to 2-in. pieces, all the fines being left in and sufficient fine material added to fill the voids. The stone was heated and mixed in pans or kettles from a street paving outfit; and the asphaltum paste, composed of 4 parts California refined asphaltum and 1 part crude petroleum, was boiled in another kettle. The boiling hot paste was poured with ladles over the hot stone, and the whole mixed over the fire with shovels and hoes. The asphalt concrete was taken away in hot iron wheelbarrows, placed in a 4-in. layer rammed and ironed with hot irons. The concrete was laid in strips 4 to 6 ft. wide, the edges being coated with hot paste. After the whole reservoir was lined, it was painted with the asphalt paste, boiled much longer, until when cold it was hard and brittle, breaking like glass under the hammer. This paste was put on very hot and ironed down. It should not be more than ⅛-in. thick or it will "creep" on slopes of 1½ to 1. After two hot summers and one cold winter there was not a single crack anywhere in the lining. A mixture of sand and asphalt will creep on slopes of 1½ to 1, but asphalt concrete will not. With asphalt at $20 a ton, and labor at $2 a day, the cost was 15 cts. a sq. ft. for 4-in. asphalt concrete. On a high slope Mr. Stanton recommends making slight berms every 6 ft. to support the concrete and prevent creeping. Asphalt concrete resists the wear of wind and water that cuts away granite and iron.

Base for Mill Floor.—In constructing 17,784 sq. ft. of tar concrete base for a mill floor, Mr. C. H. Chadsey used a sand, broken stone and tar mixture mixed in a mechanical mixer. The apparatus used and the mode of procedure followed were as follows:

Two parallel 8-in. brick walls 26 ft. long were built 4 ft. apart and 2½ ft. high to form a furnace. On these walls at one end was set a 4×6×2 ft. steel plate tar heating tank. Next to this tank for a space of 4×8 ft. the walls were spanned between with steel plates. This area was used for heating sand.

Another space of 4×8 ft. was covered with 1½ in. steel rods arranged to form a grid; this space was used for heating the broken stones. The grid proved especially efficient, as it permitted the hot air to pass up through the stones, while a small cleaning door at the ground allowed the screenings which dropped through the grid to be raked out and added to the mixture. A fire from barrel staves and refuse wood built under the tank end was sufficient to heat the tar, sand and stone.

For mixing the materials a Ransome mixer was selected for the reason that heat could be supplied to the exterior of the drum by building a wood fire underneath. This fire was maintained to prevent the mixture from adhering to the mixing blades, and it proved quite effective, though occasionally they would have to be cleaned with a chisel bar, particularly when the aggregate was not sufficiently heated before being admitted to the mixer. A little "dead oil" applied to the discharge chute and to the shovels, wheelbarrows and other tools effectually prevented the concrete from adhering to them.

The method of depositing the concrete was practically the same as that used in laying cement sidewalks. Wood strips attached to stakes driven into the ground provided templates for gaging the thickness of the base and for leveling off the surface. The wood covering consisted of a layer of 2-in. planks, covered by matched hardwood flooring. In placing the planking, the base was covered with a ¼-in. layer of hot pitch, into which the planks were pressed immediately, the last plank laid being toe-nailed to the preceding plank just enough to keep the joint tight. After a few minutes the planks adhered so firmly to the base that they could be removed only with difficulty. The hardwood surface was put on in the usual manner. The prices of materials and wages for the work were as follows:

Pitch, bulk, per lb.	$ 0.0075
Gravel, per cu. yd.	1.50
Spruce sub-floor, per M. ft. B. M.	15.00
Hardwood surface, per M. ft. B. M.	33.00
Laborers, per 10-hour day.	1.50
Foreman, per 10-hour day.	4.00
Carpenters, per 10-hour day.	2.00

At these prices and not including a small administration cost or the cost of tools and plant, the cost of the floor consisting of 4½ ins. of concrete, 2 ins. of spruce sub-flooring and ⅞-in. hardwood finish was as follows per square foot:

Pitch	$0.04
Gravel	0.02
Spruce, for sub-floor	0.03
Hardwood for surfacing	0.035
Labor, mixing	0.03
Labor, laying	0.015
Carpenter work	0.025
Total per square foot	$0.195

7

LAYING CONCRETE IN COLD WEATHER: A METHOD LOGICAL APPROACH

Reinforced concrete work may be done in freezing weather if the end to be gained warrants the extra cost. Laboratory experiments show beyond much doubt that Portland cement concrete which does not undergo freezing temperatures until final set has taken place, or which, if frozen before it has set, is allowed to complete the setting process after thawing without a second interruption by freezing, does not suffer loss of ultimate strength or durability. These requirements for safety may be satisfied by so treating the materials or compounding the mixture that freezing will not occur at normal freezing temperature or else will be delayed until the concrete has set, by so housing in the work and artificially treating the inclosed space that its temperature never falls as low as the freezing point, or, by letting the concrete freeze if it will and then by suitable protection and by artificial heating produce and maintain a thawing temperature until set has taken place.

LOWERING THE FREEZING POINT OF THE MIXING WATER.— Lowering the freezing point of the mixing water is the simplest and cheapest method by which concrete can be mixed and deposited in freezing weather. The method consists simply in adding some substance to the water which will produce a brine or emulsion that freezes at some temperature below 32° F. determined by the substance added and the richness of the admixture. A great variety of substances may be added to water to produce low freezing brines, but in concrete work only those may be used that do little or no injury to the strength and durability of the concrete. Practice has definitely determined only one of these, namely, sodium chloride or common salt, though some others have been used successfully in isolated cases. A point to be borne in mind is that cold retards the setting of cement and that the use of anti-freezing mixtures emphasizes this phenomenon and its attendant disadvantages in practical construction. The accompanying diagram, Fig. 39, based on the experiments of Tetmajer, show the effect on the freezing point of water by the admixtures of various substances that have been suggested for reducing the freezing point of mortar and concrete mixtures.

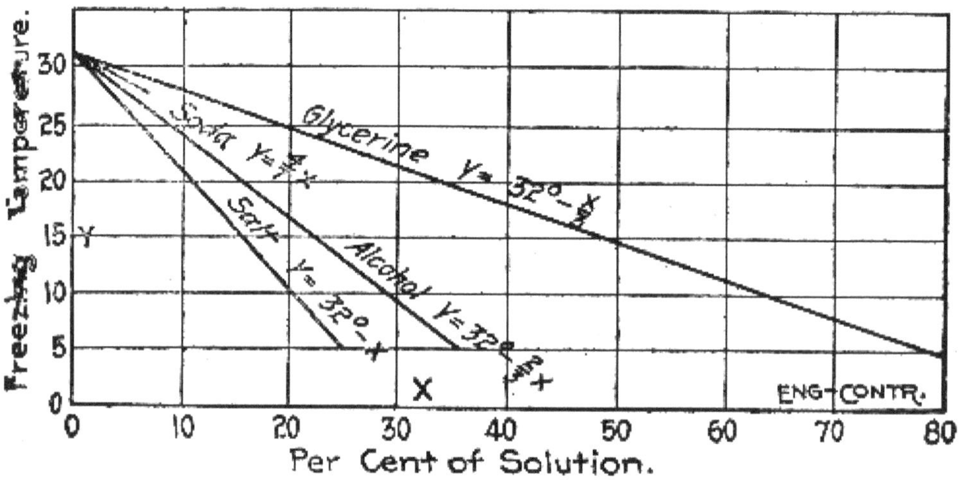

Fig. 39.—Diagram Showing Effect on Freezing Point of Water by Admixture of Various Substances.

Common Salt (Sodium Chloride).—The substance most usually employed to lower the freezing point of water used in concrete is common salt. Laboratory experiments show that the addition of salt retards the setting and probably lowers the strength of cement at short periods, but does not, when not used to excess, injure the ultimate strength. The amount beyond which the addition of salt begins to affect injuriously the strength of cement is stated variously by various authorities. Sutcliffe states that it is not safe to go beyond 7 or 8 per cent. by weight of the water; Sabin places the safe figures at 10 per cent., and the same figure is given by a number of other American experimenters. A number of rules have been formulated for varying the percentage of salt with the temperature of the atmosphere. Prof. Tetmajer's rule as stated by Prof. J. B. Johnson, is to add 1 per cent. of salt by weight of the water for each degree Fahrenheit below 32°. A rule quoted by many writers is "1 lb. of salt to 18 gallons of water for a temperature of 32° F., and an increase of 1 oz. for each degree lower temperature." This rule gives entirely inadequate amounts to be effective, the percentage by weight of the water being about 1 per cent. The familiar rules of enough salt to make a brine that will "float an egg" or "float a potato" are likewise untrustworthy; they call respectively, according to actual tests made by Mr.

Sanford E. Thompson, for 15 per cent. and 11 per cent. of salt which is too much, according to the authorities quoted above, to be used safely. In practice an arbitrary quantity of salt per barrel of cement or per 100 lbs. of water is usually chosen. Preferably the amount should be stated in terms of its percentage by weight of the water, since if stated in terms of pounds per barrel of cement the richness of the brine will vary with the richness of the concrete mixture, its composition, etc. As examples of the percentages used in practice, the following works may be quoted: New York Rapid Transit Railway, 9 per cent. by weight of the water; Foster-Armstrong Piano Works, 6 per cent. by weight of the water. In summary, it would seem that if a rule for the use of salt is to be adopted that of Tetmajer, which is to add 1 per cent. by weight of the water for each degree Fahrenheit below 32°, is as logical and accurate as any. It should, however, be accompanied by the proviso that no more than 10 per cent. by weight of salt should be considered safe practice, and that if the frost is too keen for this to avail some other method should be adopted or the work stopped. It may be taken that each unit per cent. of salt added to water reduces the freezing temperature of the brine about 1.08° F.; a 10 per cent. salt brine will therefore freeze at 32° - 11° = 21° F. The range of efficiency of salt as a preventative of frost in mixing and laying concrete is, obviously, quite limited.

HEATING CONCRETE MATERIALS.

mixing water acts both to hasten the setting and to lengthen the time before the mixture becomes cold enough to freeze. At temperatures not greatly below freezing the combined effects are sufficient to ensure the setting of the concrete before it can freeze. More specific data of efficiency are difficult to arrive at. There are no test data that show how long it takes a concrete mixture at a certain temperature to lose its heat and become cold enough to freeze at any specific temperature of the surrounding air, and a theoretical calculation of this period is so beset with difficulties as to be impracticable. Strength tests of concrete made with heated materials have shown clearly enough that the heating has no effect worth mentioning on either strength or durability. Either the water, the sand, the aggregate or all three may be heated; usually the cement is not heated but it may be if desired.

Portable Heaters.—An ordinary half cylinder of sheet steel set on the ground like an arch is the simplest form of sand heater. A wood fire is built under the arch and the sand to be heated is heaped on the top and sides. The efficiency of this device may be improved by closing one end of the arch and adding a short chimney stack, but even the very crude arrangement of of corrugated iron bent to an arc will do good service where the ies handled are small. This form of heater may be used for stone or in the same manner as for sand. It is inexpensive, simple to operate quires only waste wood for fuel, but unless it is fired with exceeding e sand in contact with the metal will be burned. The drawings of Fig. w the construction of a portable heater for sand, stone and water used structing concrete culverts on the New York Central & Hudson River Railroad. This device weighs 1,200 lbs., and costs about $50.

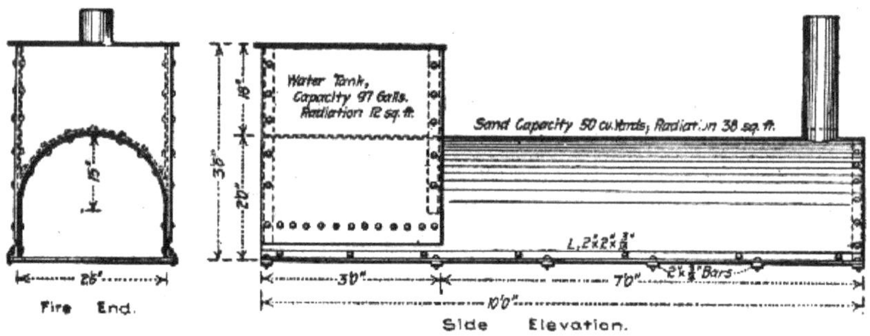

Fig. 40.—Portable Sand, Stone and Water Heater.

Heating in Stationary Bins.—The following arrangement for heating sand and gravel in large quantities in bins was employed in constructing the Foster-Armstrong Piano Works at Rochester, N. Y. The daily consumption of sand and gravel on this work was about 50 cu. yds. and 100 cu. yds., respectively. To provide storage for the sand and gravel, a bin 16 ft. square in projected plan was constructed with vertical sides and a sloping bottom as illustrated in Fig. 41. This bin was divided by a vertical partition into a large compartment for gravel and a small compartment for sand and was provided with two grates of boiler tubes arranged as shown. These grates caused V-shaped cavities to be formed beneath in the gravel and sand. Into these cavities penetrated through one end of the bin 6-in. pipes from a hot air furnace and 1-in. pipes from a steam boiler. The hot air pipes merely pass through the wall but the steam pipes continue nearly to the opposite side of the bin and are provided with open crosses at intervals along their length. In addition to the conduits described there is a small pipe for steam located below and near the bottom of the bin. The hot air pipes connected with a small furnace and air was forced through them by a Sturtevant No. 6 blower. The steam pipes connected with the boiler of a steam heating system installed to keep the buildings warm during construction.

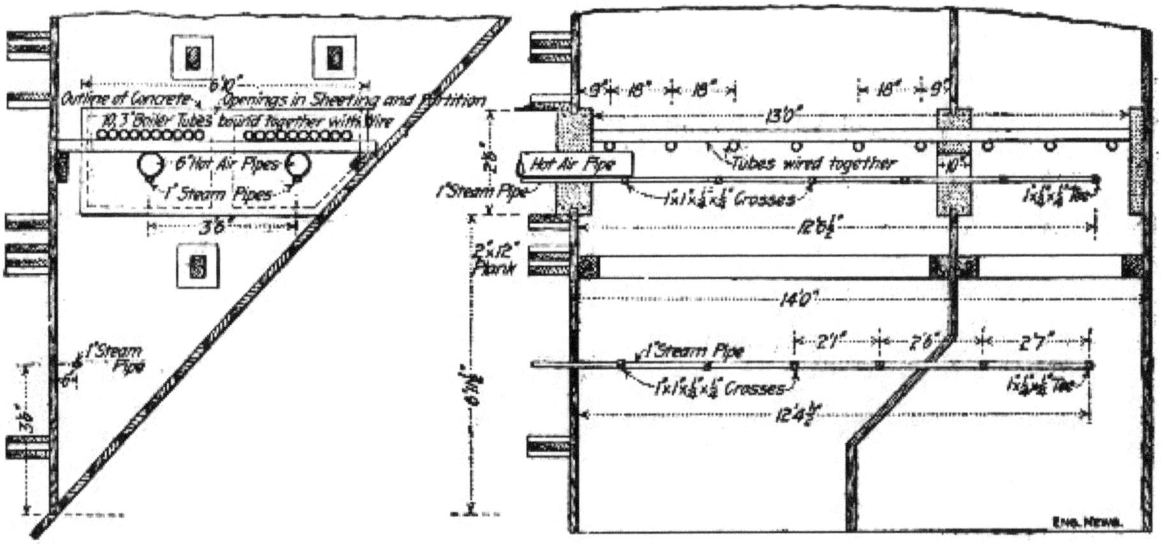

Fig. 41.—Bin Arrangement for Heating Sand and Stone.

Other Examples of Heating Materials.—In the construction of the power plant of the Billings (Mont.) Water Power Co., practically all of the concrete work above the main floor level was put in during weather so cold that it was necessary to heat both the gravel and water used. A sand heater was constructed of four 15-ft. lengths of 15-in. cast iron pipe, two in series and the two sets placed side by side. This gave a total length of 30 ft. for heating, making it possible to use the gravel from alternate ends and rendering the heating process continuous. The gravel was dumped directly on the heater, thus avoiding the additional expense of handling it a second time. The heater pipes were laid somewhat slanting, the fire being built in the lower end. A 10-ft. flue furnished sufficient draft for all occasions. With this arrangement it was possible to heat the gravel to a temperature of 80° or 90° F. even during the coldest weather. Steam for heating the water was available from the plant. The temperature at which the concrete was placed in the forms was kept between 65° and 75° F. This was regulated by the man on the mixer platform by varying the temperature of the water to suit the conditions of the gravel. When the ingredients were heated in this manner it was found advisable to mix the concrete "sloppy," using even more water than would be commonly used in the so-called "sloppy" concrete. No difficulty was experienced with temperature cracks if the concrete, when placed, was not above 75° F. All cracks of this nature which did appear were of no consequence, as they never extended more than ½ in. below the surface. The concrete was placed in as large masses as possible. It was covered nights with sacks and canvas and, when the walls were less than 3 ft. in width, the outside of the forms was lagged with tar paper. An air space was always left between the surface of the concrete and the covering. Under these conditions there was sufficient heat in the mass to prevent its freezing for several days, which was ample time for permanent setting.

During the construction in 1902 of the Wachusett Dam at Clinton, Mass., for the Metropolitan Water Works Commission the following procedures were followed in laying concrete in freezing weather: After November 15 all masonry was laid in Portland cement, and after November 28 the sand and water were heated and salt added in the proportion of 4 lbs. per barrel of cement. The sand was heated in a bin, 16½×15½×10 ft. deep, provided with about 20 coils of 2-in. pipe, passing around the inside of the bin. The sand, which was dumped in the top of the bin and drawn from the bottom, remained there long enough to become warm. The salt for each batch of mortar was dissolved in the water which was heated by steam; steam was also used to thaw ice from the stone masonry. The laying of masonry was not started on mornings when the temperature was lower than 18° F. above zero, and not even with this temperature unless the day was clear and higher temperature expected. At the close of each day the masonry built was covered with canvas.

In the construction of dams for Huronian Company's power development in Canada a large part of the concrete work in dams, and also in power house foundations, was done in winter, with the temperature varying from a few degrees of frost to 15 degrees below zero, and on several occasions much lower. No difficulty was found in securing good concrete work, the only precaution taken being to heat the mixing water by turning a ¾-in. steam pipe into the water barrel supplying the mixer, and, during the process of mixing, to use a jet of live steam in the mixer, keeping the cylinder closed by wooden coverings during the process of mixing. No attempt was made to heat sand or stone. In all the winter work care was taken to use only cement which would attain its initial set in not more than 65 minutes.

In constructing a concrete arch bridge at Plano, Ill., the sand and gravel were heated previous to mixing and the mixed concrete after placing was kept from freezing by playing a steam jet from a hose connected with the boiler of the mixer on the surface of the concrete until it was certain that initial set had taken place. Readings taken with thermometers showed that in no instance did the temperature of the concrete fall below 32° F. within a period of 10 or 12 hours after placing.

From experience gained in doing miscellaneous railway work in cold weather Mr. L. J. Hotchkiss gives the following:

"For thin reinforced walls, it is not safe to rely on heating the water alone or even the water and sand, but the stone also must be heated and the concrete when it goes into the forms should be steaming hot. For mass walls the stone need not be heated except in very cold weather. Where concrete is mixed in small quantities the water can be heated by a wood fire, and if a wood fire be kept burning over night on top of the piles of stone and sand a considerable quantity can be heated. The fire can be kept going during the day and moved back on the pile as the heated material is used. This plan requires a quantity of fuel which in most cases is prohibitive and is not sufficient to supply a power mixer. For general use steam is far better.

"A convenient method is to build a long wooden box 8 or 10 in. square with numerous holes bored in its sides. This is laid on the ground, connected with a steam pipe and covered with sand, stone or gravel. The steam escaping through the holes in the box will heat over night a pile of sand, or sand and gravel, 8 or 10 ft. high. Perforated pipes can be substituted for boxes. Material can be heated more rapidly if the steam be allowed to escape in the pile than if it is confined in pipes which are not perforated. Crushed stone requires much more heat than sand or sand and gravel mixed because of the greater volume of air spaces. In many cases material which has already been unloaded must be heated. The expense of putting steam boxes or pipes under it is considerable. To avoid this one or more steam jets may be used, the end of the jet pipe being pushed several feet into the pile of material. If the jets are connected up with steam hose they are easily moved from place to place. It is difficult to heat stone in this way except in moderate weather.

"On mass work and at such temperatures as are met with in this latitude (Chicago, Ill.) it is not usually necessary to protect concrete which has been placed hot except in the top of the form. This can be done by covering the top of the form with canvas and running a jet of steam under it. If canvas is not available boards and straw or manure answer the purpose. If heat is kept on for 36 hours after completion, this is sufficient, except in unusually cold weather. The above treatment is all that is required for reinforced retaining walls of ordinary height. But where box culverts or arches carrying heavy loads must be placed in service as soon as possible, the only safe way is to keep the main part of the structure warm until the concrete is thoroughly hardened. Forms for these structures can be closed at the ends and stoves or salamanders kept going inside, or steam heat may be used. The outside may be covered with canvas or boards, and straw and steam jets run underneath. After the concrete has set enough to permit the removal of the outer forms of box culverts, fires may be built near the side walls and the concrete seasoned rapidly. Where structures need not be loaded until after the arrival of warm weather, heat may be applied for 36 hours, and the centering left in place until the concrete has hardened. Careful inspection of winter concrete should be made before loads are applied. In this connection it may be noted that concrete which has been partly seasoned and then frozen, closely resembles thoroughly seasoned concrete. Pieces broken off with a smooth fracture through all the stones and showing no frost marks, when thawed out, can be broken with the hands."

In building Portland cement concrete foundations for the West End St. Ry., Boston, and the Brooklyn Heights R. R., much of the work was done in winter. A large watertight tank was constructed, of such size that three skips or boxes of stone could be lowered into it. The tank was filled with water, and a jet of steam kept the water hot in the coldest weather. The broken stone was heated through to the temperature of the water in a few minutes. One of the stone boxes was then hoisted out, and dumped on one side of the mixing machine, and then run through the machine with sand, cement and water. The concrete was wheeled to place without delay and rammed in 12-in. layers. The heat was retained until the cement was set. In severely cold weather the sand was heated and the mixing water also. A covering of hay or gunnysacks may be used.

COVERING AND HOUSING THE WORK.—Methods of covering concrete to protect it from light frosts such as may occur over night will suggest themselves to all; sacking, shavings, straw, etc., may all be used. Covering wall forms with tar paper nailed to the studding so as to form with the lagging a cellular covering is an excellent device and will serve in very cold weather if the sand and stone have been heated. From these simple precautions the methods used may range to the elaborate systems of housing described in the following paragraphs.

Method of Housing in Dam, Chaudiere Falls, Quebec.—In constructing a dam for the water power plant at Chaudiere Falls, P. Q., the work was housed in. The wing dam and its end piers aggregated about 250 ft. in length by about 20 ft. in width. A house 100 ft. long and 24 ft. wide was constructed in sections about 10 ft. square connected by cleats with bolts and nuts. This house was put up over the wing dam. It was 20 ft. high to the eaves, with a pitched roof, and the ends were closed up; in the roof on the forebay side were hatchways with sliding doors along the whole length. Small entrance doors for the workmen were provided in the ends of the building. The house was heated by a number of cylindrical sheet-iron stoves about 18 ins. in diameter by 24 ins. high, burning coke; thermometers placed at different points in the shed gave warning to stop work when the temperature fell below freezing, which, however, rarely occurred. Mixing boards were located in the shed, and concrete, sand and broken stone were supplied in skipfuls by guy derricks located in the forebay, which passed the material through the hatchways in the roof, the proper hatchway being opened for the purpose and quickly closed. The mortar was first mixed on a board, and then a skip-load of stone was dumped into the middle of the batch and the whole well mixed. The water was made lukewarm by introducing a steam-jet into several casks which were kept full. The sand was heated outside in the forebay on an ordinary sand heater. The broken stone was heated in piles by a steam-jet; a pipe line on the ground was made up of short lengths of straight pipe alternating with T-sections—turned up. The stone was piled 3 to 4 ft. deep over the pipe and a little steam turned into the pipe. Several such piles kept going all the time supplied enough stone for the work; the stone was never overheated, and was moist enough not to dry out the mortar when mixed with it. In this manner the concreting was successfully carried on and the wing dam built high enough to keep high water out of the forebay.

Some danger from freezing was also encountered the next season, when the last part of the wing dam was being constructed. This work was done when the temperature was close to freezing, and it became necessary to keep the freshly placed concrete warm over night. This was done by covering the work loosely with canvas, under which the nozzle of a steam hose was introduced. By keeping a little steam going all night the concrete was easily kept above freezing temperature.

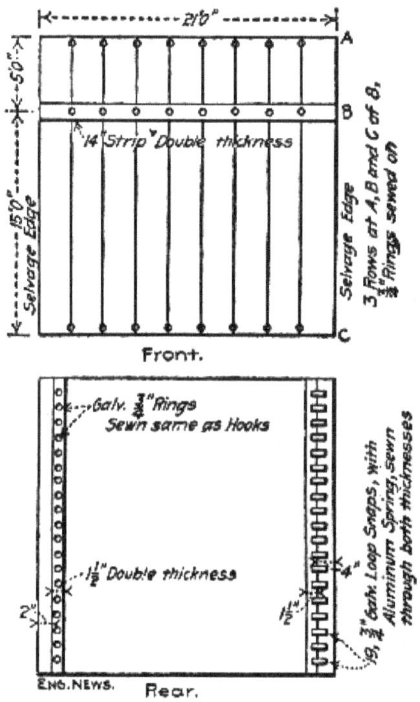

Fig. 42.—Canvas Curtain for Enclosing Open Walls.

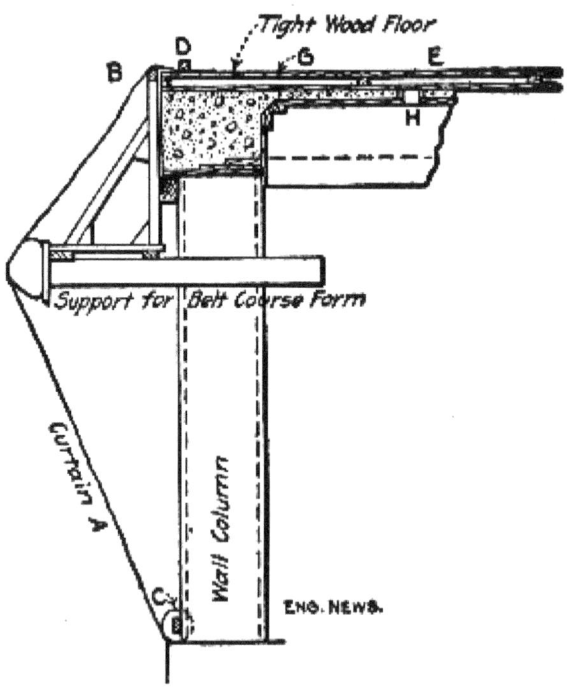

Fig. 43.—Sketch Showing Method of Applying Curtains to Open Walls.

Method of Housing in Building Work.—The following method of housing in building work is used by Mr. E. L. Ransome. The feature of the system is that the enclosing structure is made up of a combination of portable units which can be used over and over again in different jobs. The construction is best explained in connection with sketches.

Figure 43 shows a first floor wall column with the wall girder surmounting it and the connecting floor system. It will be seen that the open sides are enclosed by canvas curtains and the floor slab is covered with wood shutters. The curtains are composed of separate pieces so devised that they may be attached to each other by means of snaps and eyes; one of these curtain units is shown by Fig. 42. Referring now to Fig. 43, the curtain *A* is held by the tying-rings to a continuous string piece *B*, the upper portion or flap *D* being held down by a metal bar or other heavy object so as to lap over the floor covers *E*. The lower edge of the curtain is attached to the string piece *C*. The sketch has been made to show how the curtain adjusts itself to irregular projections such as the supports for a wall girder form; to prevent the curtain tearing on such projections it is well to cover or wrap the rough edges with burlap, bagging or other convenient material. The details of the wooden floor covers are shown by Fig. 44; they are constructed so as to give a hollow space between them and the floor and holes are left in the floor slab as at *H*, Fig. 43, to permit the warm air from below to enter this hollow space. This warm air is provided by heating the enclosed story of the building by any convenient adequate means. In constructing factory buildings, 50×200 ft. in plan at Rochester, N. Y., Mr Ransome used a line of ¾ to ⅜-in. steam pipe located at floor level and running around all four sides and a similar line running lengthwise of the building at the center, these pipes discharging live steam through openings into the enclosed space. In addition to the steam piping 10 braziers in which coke fires were kept were scattered around the floor. This equipment kept the enclosed story, 50×100 ft.×13 ft. high, at a temperature of 80° F. and at temperature of about 40° F. between the floor top and its board covering. The work was not stopped at any time because of cold and the temperatures outside ranged from zero to 10° above.

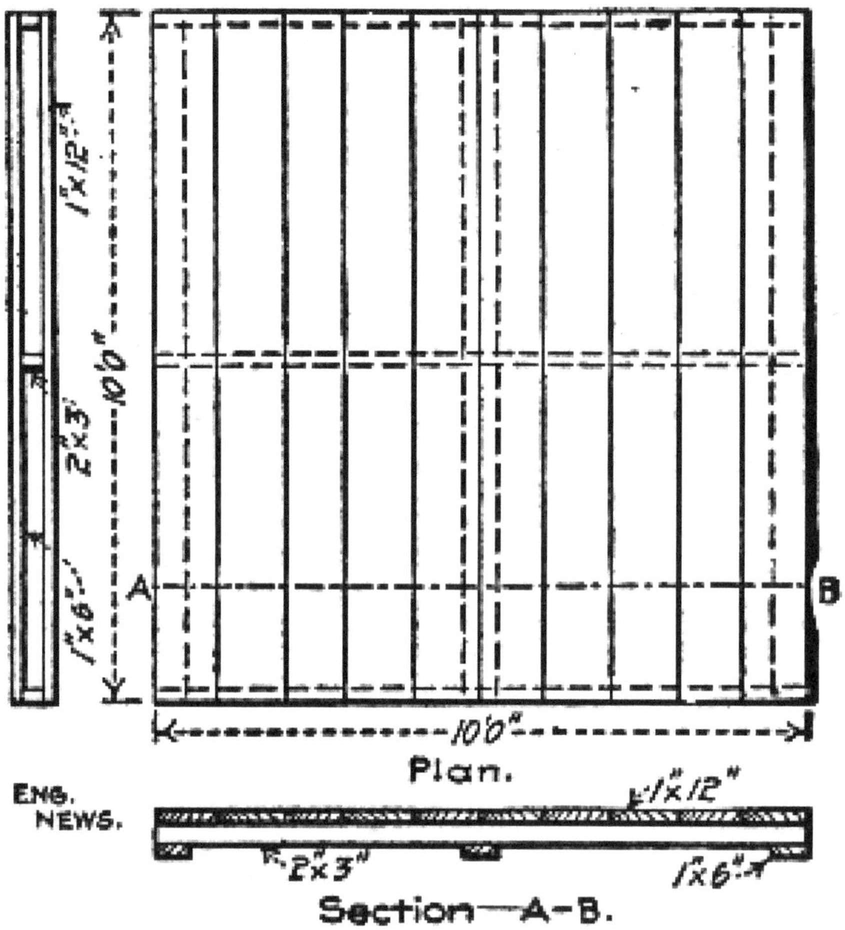

Fig. 44.—Portable Wooden Panels for Covering Floors.

8

CONCRETE SURFACES: FINISHING TOUCHES

Good design in concrete as well as in steel, masonry and wood, requires that the structure shall be good to look at. This means that the proportions must be good and that the surface finish must be pleasing. Good proportions are a matter of design but a pleasing surface finish is a matter of construction. Many, perhaps the majority of, concrete structures do not have a pleasing surface finish; the surface is irregular, uneven in texture, and stained or discolored or of lifeless hue. The reasons for these faults and the possible means of remedying them are matters that concern the construction engineer and the contractor.

Imperfections in the surface of concrete are due to one or more of the following causes: (1) Imperfectly made forms; (2) imperfectly mixed concrete; (3) carelessly placed concrete; (4) use of forms with dirt or cement adhering to the boards; (5) efflorescence and discoloration of the surface after the forms are removed.

IMPERFECTLY MADE FORMS.—In well mixed and placed concrete the film of cement paste which flushes to the surface will take the impress of every flaw in the surface of the forms. It will even show the grain marks in well dressed lumber. From this it will be seen how very difficult it is so to mold concrete that the surface will not bear evidence of the mold used. The task is impracticable of perfect accomplishment and the degree of perfection to which it can be carried depends upon the workmanship expended in form construction. Forms with a smooth and even surface are difficult and expensive to secure. It is impracticable in the first place to secure lagging boards dressed to exact thickness and in the second place it is impracticable to secure perfect carpenter work; joints cannot be got perfectly close and a nail omitted here or there leaves a board free to warp. From this point on the use of imperfectly sized lumber and careless carpentry can go to almost any degree of roughness in the form work. Only approximately smooth and unmarked concrete surfaces can be secured in plain wooden forms and this only with the very best kind of form construction. So much for the limitations of form work in the matter of securing surface finish. These limitations may be reduced in various ways. Joint marks may be eliminated wholly or partly by pointing the joints with clay or mortar or by pasting strips of paper or cloth over them, or the whole surface of the lagging can be papered; by the use of oiled paper there will be little trouble from the paper sticking. Grain marks and surface imperfections can be reduced by oiling the lumber so as to fill the pores or by first oiling and then filling the coat of oil with fine sand blown or cast against the boards.

The preceding remarks are of course based on the assumption that as nearly as possible a smooth and even surface finish is desired. If something less than this is sufficient, and in many cases it is, form produced surface defects become negligible in the proportion that they do not exceed the standards of roughness and irregularity considered permissible by the engineer and these standards are individual with the engineer; what one considers excessive roughness and irregularity another may consider as amply even and smooth. The point to be kept in mind is that beyond a certain state of evenness and regularity form produced surfaces are impracticable to obtain, because to construct forms of the necessary perfection to obtain them costs far more than it does to employ special supplementary finishing processes.

Surface blemishes due to dirt or cement adhering to the form boards have no excuse if the engineer or contractor cares to avoid them. It is a simple matter to keep the lagging clean and free from such accumulations.

IMPERFECT MIXING AND PLACING.—Imperfectly mixed and placed concrete gives irregularly colored, pitted and honeycombed surfaces with here a patch of smooth mortar and there a patch of exposed stone. Careful mixing and placing will avoid this defect, or all chance of it may be eliminated by using surface coatings of special mixtures. There is no great difficulty, however, in obtaining a reasonably homogeneous surface with concrete; the task merely requires that the mixing shall be reasonably uniform and homogeneous and that in placing this mixture the spading next to the lagging shall be done in such a way as to pull the coarse stones back and flush the mortar to the surface. Spading forks are excellent for this purpose. A better tool is a special spade made with a perforated blade; this special spade costs $3.

EFFLORESCENCE.—Efflorescence is the term applied to the whitish or yellowish accumulations which often appear on concrete surfaces. "Whitewash" is another name given to these blotches. Efflorescence is due to certain salts leaching out of the concrete and accumulating into thin layers where the water evaporates on the surface. These salts are most probably sulphates of calcium and magnesium, both of which are contained in many cements and both of which are slightly soluble in water. Efflorescence is very erratic in its appearance. Some concretes never exhibit it; in some it may not appear for several years, and in others it shows soon after construction and may appear in great quantities. The most effective way to prevent efflorescence would naturally be to use cements entirely free from sulphates, chlorides or whatever other soluble salts are the cause of the phenomenon, but the likelihood of engineers resorting to the trouble of such selection, except in rare instances, is not great, even if they knew what cements to select, so that other means must be sought. The most common place for efflorescence to appear in walls is at the horizontal junction of two days' work or where a coping is placed after the main body of the wall has been completed. The reason of this seems to be that the salt solutions seep down through the concrete until they strike the nearly impervious film of cement that forms on the top surface of the old concrete before the new is added, and then they follow along this impervious film to the face of the wall. The authors have suggested that this cause might be remedied by ending the day's work by a layer whose top has a slight slope down toward the rear of the wall or perhaps by placing all the concrete in similarly sloping layers. Mr. C. H. Cartlidge is authority for the statement that this leaching at joints can be largely done away with by the simple process of washing the top surface of concrete which has been allowed to set over night by scrubbing it with wire brushes in conjunction with thorough flushing with a hose. But efflorescence frequently appears on the faces of walls built without construction joints and in which a wet concrete is puddled in and not tamped in layers, and here other means are obviously essential. Waterproofing the surface of the wall should be effective so long as the waterproofing lasts; indeed one of the claims made for some of these waterproofing compounds is that efflorescence is prevented. The various waterproofing mixtures capable of such use will be found described in Chapter XXV. Failing in any or all of these methods of preventing efflorescence the engineer must resort to remedial measures. The saline coating must be scraped, or chipped, or better, washed away with acids.

Efflorescence was removed from a concrete bridge at Washington, D. C, by using hydrochloric (muriatic) acid and common scrubbing brushes; 30 gals. of acid and 36 scrubbing brushes were used to clean 250 sq. yds. of concrete. The acid was diluted with 4 or 5 parts water to 1 of acid; water was constantly played with a hose on the concrete while being cleaned to prevent penetration of the acid. One house-front cleaner and 5 laborers were employed, and the total cost was $150, or 60 cts. per sq. yd. This high cost was due to the difficulty of cleaning the balustrades. It is thought that the cost of cleaning the spandrels and wing walls did not exceed 20 cts. per sq. yd. The cleaning was perfectly satisfactory. An experiment was made with wire brushes without acid, but the cost was $2.40 per sq. yd. The flour removed by the wire brushes was found by analysis to be silicate of lime. Acetic acid was tried in place of muriatic, but required more scrubbing.

SPADED AND TROWELED FINISHES.—With wet-concrete and ordinarily good form construction a reasonably good surface appearance can be obtained by spading and troweling. For doing the spading a common gardener's hoe, straightened out so that the blade is nearly in line with the handle will do good work. The blade of the tool is pushed down next to the lagging and the stone pulled back giving the grout opportunity to flush to the face. Troweling, that is troweling without grout wash, requires, of course, that the concrete be stripped before it has become too hard to be worked. As troweling is seldom required except for tops of copings and corners it is generally practicable to bare the concrete while it is still fairly green. In this condition the edges of copings, etc., can be rounded by edging tools such as cement sidewalk workers use.

PLASTER AND STUCCO FINISH.—The ordinary concrete surface with a film-like cement covering will not hold plaster or stucco. To get proper adhesion the concrete surface must be scrubbed, treated with acid or tooled before the plaster or stucco is applied and this makes an expensive finish since either of the preliminary treatments constitutes a good finish by itself. When a coarse grained facing is made of very dry mixtures, as described in a succeeding section, it has been made to hold plaster very well on inside work. In general plaster and stucco finishes can be classed as uncertain even when the concrete surface has been prepared to take them, and when the concrete has not been so prepared such finishes can be classed as absolutely unreliable.

MORTAR AND CEMENT FACING.—Where a surface finish of fine texture or of some special color or composition is desired the concrete is often faced with a coat of mortar or, sometimes, neat cement paste or grout. Mortar facing is laid from 1 to 2 ins. thick, usually 1½-ins., the mortar being a 1-1, 1-2 or 1-3 mixture and of cement and ordinary sand where no special color or texture is sought. This facing often receives a future special finish as described in succeeding sections, but it is more usually used as left by the forms or at best with only a troweling or brushing with grout. Engineers nearly always require that the mortar facing and the concrete backing shall be constructed simultaneously. This is accomplished by using facing forms, two kinds of which are shown by Figs. 45 and 46. In use the sheet steel plates are placed on edge the proper distance back of the lagging and the space between them and the lagging is filled with the facing mortar. The concrete backing is then filled in to the height of the plate, which is then lifted vertically and the backing and facing thoroughly bonded by tamping them together. The form shown by Fig. 46, though somewhat the more expensive, is the preferable one, since the attached ribs keep the plate its exact distance from the lagging without any watching by the men, while the flare at the top facilitates filling. The facing mortar has to be rather carefully mixed; it must be wet enough to work easily and completely into the narrow space and yet not be so soft that in tamping the backing the stones are easily forced through it. Also since the facing cannot proceed faster than the backing the mortar has to be mixed in small batches so that it is always fresh. A cubic yard of mortar will make 216 sq. ft. of 1½-in a facing. Cement facing is seldom made more than 1 in. thick. If placed as paste the process is essentially the same as for placing mortar. When grout is used a form is not used; place and tamp the concrete in 6 to 8-in layers, then shove a common gardener's spade down between the concrete and the lagging and pull back the concrete about an inch and pour the opening full of grout and withdraw the spade. If this work is carefully done there will be very few stones showing when the forms are removed. When stiff pastes or mortars are used the contractor often places the facing by plastering the lagging just ahead of the concreting; this process requires constant watching to see that the plaster coat does not slough or peel off before it is backed up with concrete.

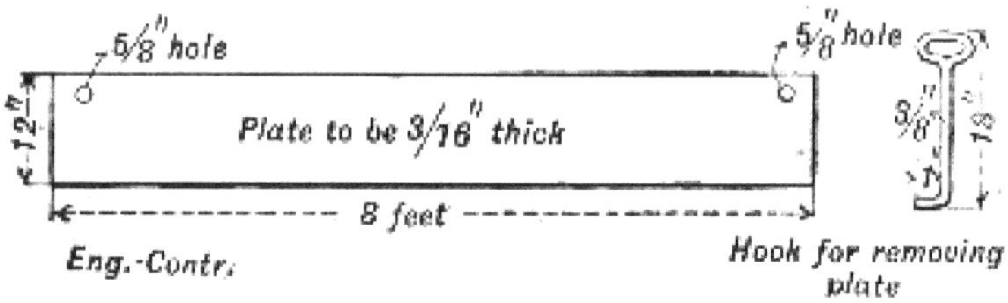

Fig. 45.—Form for Applying Cement Facing (Massachusetts Highway Commission).

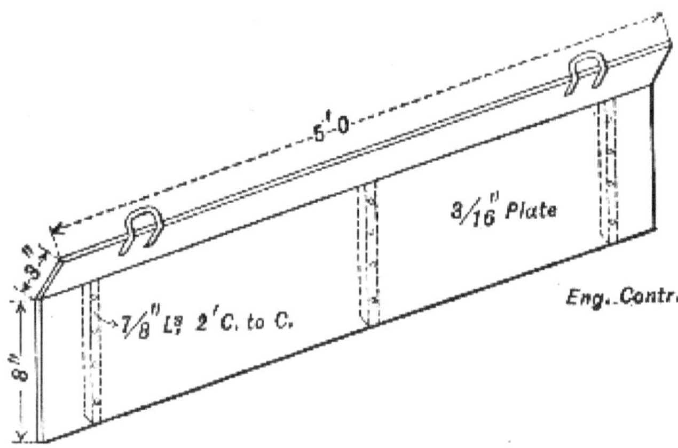

Fig. 46.—Form for Applying Cement Facing (Illinois Central R. R.).

SPECIAL FACING MIXTURES FOR MINIMIZING FORM MARKS.—The ordinary facing mixture of mortar or cement is so fine grained and plastic that it readily takes the impress of every irregularity in the form lagging; where a particularly good finish is desired this makes necessary subsequent finishing treatments. To avoid these subsequent treatments and at the same time to reduce the form marks, special facing mixtures, which will not take the imprint of and which will minimize rather than exaggerate every imperfection in the forms, have been used with very considerable success in the concrete work done for the various Chicago, Ill., parks. The mixture used consists usually of 1 part cement, 3 parts fine limestone screenings and 3 parts ¾-in. crushed limestone; these materials are mixed quite dry so no mortar will flush to the surface when rammed hard. With moderately good form work the imprint of the joints is hardly noticeable and grain marks do not show at all. For thin building walls the special mixture is used throughout the wall, but for more massive structures it is used only for the facing.

GROUT WASHES.—Grout finishes serve only to fill the small pits and pores in the surface coating; cavities or joint lines, if any exist, must be removed by plastering or rubbing before the grout is applied or else by applying the grout by rubbing. In ordinary work the grout is applied with a brush after the holes have been plastered and the joint marks rubbed down. The grout to be applied with a brush should be about the consistency of whitewash; a 1 cement 2 sand mixture is a good one. Where a more perfect finish of dark color is desired the grout of neat cement and lampblack in equal parts may be applied as follows: Two coats with a brush, the second coat after the first has dried, and one coat by sweeping with a small broom. The broom marks give a slightly rough surface. Instead of a liquid grout a stiff grout or semi-liquid mortar applied with a trowel or float can be used. In this case the grout should be applied in a very thin coat and troweled or floated so that only the pores are filled and no body of mortar left on the surface or else it will scale off. A more expensive but very superior grout finish is obtained by rubbing and scouring the wet grout into the surface with cement mortar bricks, carborundum bricks, or such like abrasive materials. A 1 cement 1 sand mortar brick, with a handle molded into it, and having about the dimensions of an ordinary building brick makes a good tool for rubbing down joint marks as well as for applying grout.

Fig. 47.—Concrete Baluster Finished by Scrubbing and Washing.

FINISHING BY SCRUBBING AND WASHING.—A successful finish for concrete structures consists in removing the forms while the concrete is green and then scrubbing the surface with a brush and water until the film of cement is removed and the clean sand or stone left exposed. This method has been chiefly used in concrete work done by the city of Philadelphia, Pa., Mr. Henry M. Quimby, Bridge Engineer. Figure 47 shows an example of scrubbed finish, but of course the texture or color of the surface will vary with the character of the face mixture and the hue of the sand or chips used. Warm tones can be secured by the use of crushed brick or red gravel; a dark colored stone with light sand gives a color much resembling granite; fine gravel or coarse sand gives a texture like sandstone. In much of this work done in Philadelphia a mixture composed of 1 part cement, 2 parts bank sand and 3 parts crushed and cleaned black, slaty shale from ⅜ to ¼ in. in size, has been used with good results both in appearance and in durability. The scrubbing is done with an ordinary house scrubbing brush at the same time flushing the concrete with water from a sponge or bucket or, preferably, from a hose. In general the washing is done on the day following the placing of the concrete but the proper time depends upon the rapidity with which the concrete sets. In warm weather 24 hours after placing is generally about right, but in cold weather 48 hours may be required, and in very cold weather the concrete has been left to set a week and the scrubbing has been successful. With the concrete in just the proper condition a few turns of the brush with plenty of water will clean away the cement, but if a little too hard wire brushes must be used and if still harder a scouring brick or an ordinary brick with sand is necessary to cut the cement film. The process requires that the forms shall be so constructed that the lagging can be removed when the concrete has reached the proper age for treatment. Mr. Quimby sets the studs 8 to 12 ins. from the face and braces the lagging boards against them by cleats nailed so as to be easily loosened. His practice is to use boards in one width the full depth of the course and to nail a triangular bead strip to the face at each edge. These bead strips mark the joints between courses as shown by Fig. 48. When a "board" is taken off it is cleaned and oiled and reset for a new course by inserting the bottom bead strip in the half indentation left by the top bead in the concrete. This is, of course, for work of such size that one course is a day's work of concreting. In such work, two carpenters with perhaps one helper will remove a course of "boards" say 100 ft. long in from 4 to 8 hours. While forms of the kind described cost more to construct there is a saving by repeated re-use of the lagging boards. The indentations or beads marking the courses serve perfectly to conceal the construction joints. The cost of scrubbing varies with the hardness of the concrete; when in just the right condition for effective work one man can scrub 100 sq. ft. in an hour; on the other hand it has taken one man a whole day to scrub and scour the same area when the concrete was allowed to get hard.

Fig. 48.—Concrete Abutment with Scrubbed Finish and Course Marks.

FINISHING BY ETCHING WITH ACID.—The acid etched or acid wash process of finishing concrete consists in first washing the surface with an acid preparation to remove the surface cement and expose the sand and stone, then with an alkaline solution to remove all free acid, and finally, with clear water in sufficient volume to cleanse and flush the surface thoroughly. The work can be done at any time after the forms are removed and does not require skilled labor; any man with enough judgment to determine when the etching has progressed far enough can do the work. This process has been very extensively used in Chicago by the South Park Commission, Mr. Linn White, Engineer. In this work the concrete is faced with a mixture of cement, sand and stone chips, any stone being used that is not affected by acid. Limestone is excluded. Where some color is desired the facing can be mixed with mineral pigments or with colored sand or stone chips. This acid wash process has been patented, the patentees being represented by Mr. J. K. Irvine, Sioux City, Ia.

TOOLING CONCRETE SURFACES.—Concrete surfaces may be bush-hammered or otherwise tool finished like natural stone, exactly the same methods and tools being used. Tooling must wait, however, until the concrete has become fairly hard. As the result of his experience in tooling some 43,000 sq. ft. of concrete, Mr. W. J. Douglas states that the concrete should be at least 30 days old and, preferably, 60 days old, if possible, when bush-hammered. There is a great variation in the costs given for tooling concrete. Mr. C. R. Neher states that a concrete face can be bush-hammered by an ordinary laborer at the rate of 100 sq. ft. in 10 hours or at a cost of 1½ cts. per square foot with wages at 15 cts. per hour. Mr. E. L. Ransome states that bush-hammering costs from 1½ to 2½ cts. per square foot, wages of common laborers being 15 cts. per hour. The walls of the Pacific Borax Co.'s factory at Bayonne, N. J., were dressed by hand at the rate of 100 to 200 sq. ft. per man per day; using pneumatic hammer one man was able to dress from 300 to 600 sq. ft. per day. In constructing the Harvard Stadium the walls were dressed with pneumatic hammers fitted with a tool with a saw-tooth cutting blade like an ice chopper. Men timed by one of the authors on a visit to this work were dressing wall surface at the rate of 50 sq. ft. per hour, but the contractor stated that the average work per man per day was 200 sq. ft. Common laborers were employed. The average cost of bush-hammering some 43,000 sq. ft. of plain and ornamental blocks for the Connecticut Avenue Bridge at Washington, D. C, was 26 cts. per square foot. Both pneumatic tools and hand tooling were employed and the work of both is lumped in the above cost, but hand tooling cost about twice as much as machine tooling. The work was done by high-priced men, foremen stone cutters at $5 per day and stone cutters at $4 per day. Moreover a grade of work equal to the best bush-hammered stone work was demanded. Full details of the cost of this work are given in Chapter XVII. Mr. H. M. Quimby states that the cost of tooling concrete runs from 3 cts. to 12 cts. per square foot, according to the character and extent of the work and the equipment.

GRAVEL OR PEBBLE SURFACE FINISH.—An effective variation of the ordinary stone concrete surface is secured by using an aggregate of rounded pebbles of nearly uniform size and by scrubbing or etching remove the cement enough to leave the pebbles about half exposed at the surface. In constructing a bridge at Washington, D. C, the concrete was a 1-2-5 gravel mixture of 1½ to 2-in. pebbles for the spandrels and arch ring face and of 1-in. pebbles for the parapet walls. The forms were removed while the concrete was still green and the cement scrubbed from around the faces and sides of the pebbles using wire brushes and water. Tests showed that at 12 hours age the concrete was not hard enough to prevent the pebbles from being brushed loose and that at 36 hours age it was too hard to permit the mortar to be scrubbed away without excessive labor; the best results were obtained when the concrete was about 24 hours old.

COLORED FACING.—Where occasion calls for concrete of a color or tint other than is obtained by the use of the ordinary materials either an aggregate of a color suitable for the purpose may be used or the mixture may be colored by the addition of some mineral pigment. The first method is by all odds the preferable one; it gives a color which will endure for all time and it in no way injures the strength or durability of the concrete. Mineral pigments may be secured from any of several well-known firms who make them for coloring concrete, and they may be had in almost every shade. Directions for using these colors can be had from the makers. All but a very few of these mineral colors injure the strength and durability of the concrete if used in amounts sufficient to produce the desired color and all of them fade in time. The best method of producing a colored mortar or concrete facing is to mix the cement with screenings produced by crushing a natural stone of the desired color.

9

FORM CONSTRUCTION

Concrete being a plastic material when deposited requires molds or forms to give it the shape required and to maintain it in that shape until it has hardened to sufficient strength to require no exterior support. The material used in constructing forms is wood. Beyond the use of metal molds for building blocks for sewer construction and for ornamental and a few architectural shapes, iron and steel are used in form construction only as ties and clamps to hold parts of wood forms together—except in rare instances. A discussion of form construction, therefore, is essentially a discussion of wood forms.

Before taking up this discussion, however, attention deserves to be called to the opportunities for the development of metal forms. Lumber is costly and is growing more scarce and costly all the time. A substitute which can be repeatedly used and whose durability and salvage value are great presents itself in steel if only a system of form units can be devised which is reasonably adjustable to varying conditions. Cylindrical steel column molds have been used to some extent and are discussed in Chapter XIX. In Chapter XVI we describe a steel form for side walls of a tunnel lining. In some building work done in the northwest corrugated steel panels or sheets have been used as lagging for floor slab centers. A number of styles of metal forms or centers for sewer and tunnel work have been devised and used and are discussed in Chapter XXI. Despite this considerable use of metal for special forms nothing approaching its general use like wood has been attempted, and the field lies wide open for invention.

The economics of form construction deserve the most serious attention of the engineer and contractor. It is seldom that form work, outside of very massive foundation construction, costs less than 50 cts. per cubic yard of concrete in place, and it is not unusual in the more complex structures for it to cost $5 per cubic yard of concrete in place. These costs include the cost of materials and of framing, handling and removing the forms but they do not embrace extremely high or low costs. It is evident without further demonstration that time spent in planning economic form construction for any considerable job of concrete work is time spent profitably.

In the following sections we review the general considerations which enter into all form work. Specific details of construction and specific costs of form work are given in succeeding chapters where each class of concrete work is discussed separately. This chapter is intended principally to familiarize the reader with general principles governing form work.

EFFECT OF DESIGN ON FORM WORK.—The designing engineer can generally aid largely in reducing the cost of form work if he will. This is particularly true in building work in which, also, form costs run high. By arranging his beam spacing and sizes with a little care he will enable the contractor to use his forms over and over and thus greatly reduce the expense for lumber. In the same way columns may be made of dimensions which will avoid frequent remaking of column forms. Panel recesses in walls may be made the thickness of a board or plank, instead of some odd depth that will require a special thickness of lumber, or beams may be made of such size that certain dimension widths of lumber can be used without splitting. In general, carpenter work costs more than concrete and where a little excess concrete may be contributed to save carpenter work it pays to contribute it. The figures given in Chapter XIX, showing the reduction in lumber cost coming from using the same material over a second or third time, should be studied in this connection. The leading firms of engineering-contractors which both design and construct reinforced concrete buildings fully realize these opportunities and take advantage of them, but the general practitioner, particularly if he be an architect, does not do so. The authors have personal knowledge of one building in which a slight change in spacing and dimensions of beams—a change that would have been of no architectural or structural significance—would have reduced the successful contractor's bid for the work by $10,000. The designing engineer should hold it as a cardinal point in design that form work, and we will add here reinforcement also, should so far as possible be made interchangeable from bay to bay and from floor to floor.

KIND OF LUMBER.—The local market and the character of the work generally determine the kind of lumber to be used for forms. The hardwoods are out of the question for form construction because they cost too much and are too hard to work. Among the soft woods white pine costs too much for general use and hemlock is unreliable when exposed to the weather. This reduces the list generally available to spruce, Norway pine and the southern pines. Neither green nor kiln-dried lumber is so good as partially dry stuff, since the kiln-dried lumber swells and crushes or bulges the joints and green lumber does not swell enough to close the joints. Forms have to withstand, temporarily, very heavy loads, therefore, knots, shakes and rot must be watched after. The choosing of good lumber is a simple process and the contractor who wants to be able to rely on his forms will look after it carefully, without going to extremes which the work does not warrant.

FINISH AND DIMENSIONS OF LUMBER.—Dressing the lumber serves four important purposes: It permits the forms to be constructed more nearly true to line and surface; it permits tighter joint construction; it gives a smoother surface finish to the concrete, and it facilitates the removal and cleaning of the forms. Undressed lumber may be used for the backs of walls and abutments, for work below ground and wherever a smooth and true surface is unimportant; there are some contractors, however, who prefer lumber dressed on one side even for these purposes because of the smaller cost of cleaning. For floor and wall forms the lumber should always be dressed on one side; where the work is very particular both sides should be dressed, and in special cases the sides of the joists or studs against which the lagging lies may be dressed. For ordinary work a square edge finish does well enough but for fine face work a tongue and groove or bevel edge finish is preferable. The tongue and groove finish gives a somewhat tighter joint on first laying but it does not take up swelling or resist wear so well as the bevel edge finish.

When ordering new lumber for forms the contractor will save much future work and waste if he does it from plans. Timber cut to length and width to go directly into the forms reduces both mill and carpenter work on the site, and in many cases it can be so ordered if ordered from plans. Waste is another item that is reduced by ordering from plans; with lumber costing its present prices crop ends run into money very rapidly. When old lumber from a previous job is to be used the contractor can only make the best of his stock, but even here form plans will result in saving. Sort and pile the old lumber according to sizes and make a schedule of the quantity of each size on hand; this schedule in the hands of the man who designs the forms and of the head carpenter will materially reduce waste and carpenter work. It is often possible especially in making concrete foundations for frame buildings to use lumber for forms which is subsequently used for floor beams, etc., in the building.

Contractors differ greatly in their ideas of the proper thickness of lumber to use for various parts of form work. Generally speaking 1¼ to 2-in. stuff is used for wall lagging held by studding and 1-in. stuff when built into panels; for floor lagging 1-in stuff with joists spaced up to 24 ins. or when built into panels; for column lagging 1¼ to 2-in. stuff; for sides of girders 1, 1¼, 1½ and 2-in. stuff are all used; and for bottoms of girders, 1½ and 2-in. stuff. These figures are by no means invariable as a study of the numerous examples of actual form work given throughout this book will show.

COMPUTATION OF FORMS.—If the minimum amount of lumber consistent with a given deflection is to be used in form work the sizes and spacing of the supporting members must be actually computed for the loading. As a practical matter of fact the amount of material used and the arrangement of the supports are often subject to requirements of unit construction, clearance, staging, etc., which supersede the matter of economical adaptation of material to loading. The designing of form work is at best, therefore, a compromise between rules of thumb and scientific calculation. In wall work empirical methods are nearly always followed. In girder and floor slab work, on the other hand, design is commonly based on computation.

In the matter of loads the general practice is to assume the weight of concrete as a liquid at some amount which it is considered will also cover the weight of men, barrows, runways and current construction materials. The assumed weights vary. One prominent engineering firm assumes the load to be the dead weight of concrete as a liquid and the load due to placing and specifies that the forms shall be designed to carry this load without deflection. Mr. W. J. Douglas, Engineer of Bridges, Washington, D. C, assumes for lateral thrust on wall forms that concrete is a liquid of half its own weight, or 75 lbs. per cu. ft. Mr. Sanford E. Thompson, Consulting Engineer, Newton Highlands, Mass., assumes for dead load, weight of concrete including reinforcement as 154 lbs. per cu. ft., and for live load, 75 lbs. per sq. ft. on slabs and 50 lbs. per sq. ft. in figuring beam and girder forms and struts.

The assumed safe stresses in form work may be taken somewhat higher than is usual in timber construction, because of the temporary character of the load. In calculating beams the safe extreme fiber stress may be assumed at 750 lbs. per sq. in. The safe stress in pounds per square inch for struts or posts is shown by Table XV, compiled by Mr. Sanford E. Thompson. The sizes of struts given are those most commonly used in form work.

Table XV.—Safe Strength of Timber Struts for Frame Work.

	—Dimensions of Strut.—			
Length Strut.	3×4-in.	4×4-in.	6×6-in.	8×8-in.
Feet.	Lbs.	Lbs.	Lbs.	Lbs.
14	.....	700	900	1,100
12	600	800	1,000	1,200
10	700	900	1,100	1,200
8	850	1,050	1,200	1,200
6	1,000	1,200	1,200	1,200

In using this table it must be borne in mind that bracing both ways reduces the length of a long strut. For example, if a strut 24 ft. long be divided into three panels by bracing the length of strut so far as the table is concerned is 8 ft.

As stated above wall forms are rarely computed. Experience has shown that the maximum spans of various thicknesses of lagging between supports are: 1-in. boards, 24 ins.; 1½-in. plank, 4 ft., and 2-in. plank, 5 ft. Studding will vary in size from 2×4 to 4×6 ins., strutted and braced horizontally to meet conditions. Column forms, like wall forms, are rarely computed, yokes being spaced 2 ft. apart for 1¼-in. lagging up to 3 to 3½ ft. apart for 2-in. lagging.

Floor forms, including girder and slab forms, are computed on the basis of a maximum deflection and not on the basis of strength. Sagging forms are liable to rupture the beam or slab. The amount of deflection considered allowable varies from no deflection up to ⅜ to ½ in. Assuming the deflection, permissible thickness of the timber necessary to carry the load is determined by the formulas:

$$d = 5\, W\, l^3 \div 384\, E\, I \quad (1)$$

and

$$I = \frac{bh^3}{12} \quad (2)$$

Formula (1) is the familiar one for computing deflection for a beam supported (not fixed) at the ends. Mr. Sanford F. Thompson suggests using the constant {3/384}, which is an approximate mean between {1/384} that for beams with fixed ends and {5/384} that for beams with ends supported. Formula (1) then becomes

$$d = 3\,W\,l^3 \div 384\,E\,I,$$

in which as above:

> d = maximum deflection in inches.
> W = total load on plank or joist.
> l = length between supports in inches.
> E = modulus of elasticity of lumber.
> I = moment of inertia of cross-section.
> b = breadth of lumber.
> h = depth of lumber.

The deflection, d, being assumed formula (1) is solved for I, moment of inertia. Substituting the value of I in formula (2) we can readily estimate the size of joist or thickness of plank to use.—For spruce, yellow pine and the other woods commonly used in form work E may be taken equal to 1,300,000 lbs. per sq. in.

DESIGN AND CONSTRUCTION.—The main points to be kept in mind in the original design and construction of forms are: Economy in lumber, economy in carpenter work, and economy in taking down, carrying and re-erecting. Economy in lumber is not merely the matter of using the least amount of lumber that will serve the purpose considering the form as an isolated structure. It may be possible to build a column form, for example, of very light material which will serve to mold a single column, but it is evident that we could better afford to use twice this amount of lumber if by doing so we obtained a form which could be used over again to mold a second column; no more lumber per column would be used while the cost of erecting a form already framed is less than the cost of framing a new form. Economy in lumber in form construction involves, therefore, recognition of the economies to be gained by repeated use of the lumber. A certain amount of additional sturdiness is required in the shape of heavier form lumber and stronger framing to provide for the wear and tear of repeated use, and it is always economy to provide it when repeated use is possible. The thing can be overdone, however; there is an economical limit to repeated use, as we demonstrate further on. In the matter of economy in carpenter work, a certain amount of extra work put into framing the forms to withstand the stress of repeated use is economically justifiable. Also carpenter work put into framing which substitutes clamps and wedges for nails is sound economy; generally speaking a skillful form carpenter is recognized by the scarcity of nails he uses. The possibility of reducing carpenter work by ordering lumber to length and width from plans has already been mentioned. It is possible often to go a step further by having certain standard panels, boxes, etc., made in regular shops. Piece work is often possible and will frequently reduce framing costs. In designing for economy in taking down, carrying and re-erecting forms a cardinal point should be that the work be such that it can be executed by common laborers. This result can be very nearly approached by careful design, even for form work that is quite complex, if a special gang is devoted to the work and trained a little in the various operations. Design the forms so that they come apart in units by simply removing bolts, clamps and wedges. They can then be taken down, carried and erected by common laborers with a skilled man in charge to meet emergencies and to true and line up the work.

In the matter of details the joints deserve particular attention. In column and girder forms, generally, joints will be square or butt joints, and to get them tight the lumber must be dressed true to edge. Tight joints are considered essential by many not only to avoid joint marks but for the more important reason that otherwise, with wet mixtures, a honeycombed concrete is produced by leakage. Where tight joints are desired tongue and groove stock or stock cut with one edge beveled and the other square give the best results. The authors believe that the best general satisfaction will be got from the bevel edge stock placed so that the bevel edge of one board comes against the square edge of the next board; undue swelling then results in the bevel edge cutting into the adjacent square edge without bulging. Tongues and grooves suffer badly from breakage. As a matter of fact square edged stock, if well dressed and sized and well filled with moisture, can be used and is used with entire success in nearly all kinds of work. The leakage will be very slight with ordinarily good butt joints and so far as surface appearance goes joint marks are more cheaply and more satisfactorily eliminated by other means than attempting to get cabinet work in form construction. Where girder forms join columns or beams connect with girders and at the angles of floor slabs with beams the edges or corners of the forms should be rounded. The edges of beams and column corners will appear better if beveled; a triangular strip in the corners of the forms will provide this bevel. Forms and mold construction for ornamental work call for and are given special consideration in Chapter XXIII. In conclusion, the reader should study the specific examples of form construction for different purposes that are given throughout the book for hints as to special practice and details.

UNIT CONSTRUCTION OF FORMS.—Unit construction has a somewhat variable meaning in form work. In wall and tank work and in some other kinds of work unit construction means the use of form units which are gradually moved ahead or upward as the concreting progresses or of form units which are used one after another in continuous succession as the concreting progresses. In column, girder and floor work unit construction means the division of the form work as a whole and also of the individual forms into independent structural units; thus in forms for a building the column forms may be independent of the girder forms and also each column and girder form be made up of several separate units. In all cases unit construction has for its purpose the use of the same form or at least the same form lumber over and over for molding purposes. Every time the use of the same form is repeated, the cost of form work per cubic yard of concrete placed is reduced. The theoretical limit of economical repetition is then the limit of endurance of the form, the practical limit, however, is something quite different. Most concrete work varies in form or dimensions often enough to prevent the use of the same forms more than a few times, and even if these variations did not exist the time element would enter to prevent the same form or form lumber being used more than a certain number of times. Unit construction to give repeated use of forming has, therefore, its economic limits. The significance of this conclusion does not lie in any novelty that it possesses but in the fact that for any piece of work it determines the labor that may profitably be expended in working out and constructing form units.

LUBRICATION OF FORMS.—All forms for concrete require a coating of some lubricant to prevent the concrete from adhering to the wood with which it comes in contact. Incidentally this coating tends to give a smoother surface to the concrete and to preserve the wood against damage by its alternate wetting and drying. The great value of lubrication is, however, that it reduces the cost of removing forms. The requisite of a good coating material is that it shall be thin enough to spread evenly and to fill the pores and grain of the wood. Crude oil or petroline makes one of the best coatings, but various other greasy substances will serve. Where the forms are not to be removed until the concrete has set hard a thorough wetting of the wood just before the concrete is placed is all the coating necessary. Any concrete adhering to forms should be thoroughly cleaned off before they are used again and the wood underneath given a special heavy coating.

FALSEWORKS AND BRACING.—The falseworks which support the forms proper and stagings for workmen, runways, material hoists, etc., do not call for any striking differences in construction and arrangement from such work elsewhere. For wall forms inclined props reaching from ground to studding are used for walls of moderate height such as retaining walls, wing walls, and abutments. For building walls of some height a gallows frame arrangement or the common braced staging used by masons and carpenters is used. In building construction, however, movable forms are commonly employed for walls more than one story high and should always be employed above one story to save staging timber. Column forms are seldom braced unless erected without connecting girder or floor forms at their tops, and then only by diagonal props to the floor or ground. Girder and floor supports usually consist of uprights set under the girder form at intervals and occasionally under floor slab forms. The spacing of props and uprights will be regulated by the judgment of the foreman and boss carpenter; no general rule is applicable, except that enough lumber must be used to hold the forms rigid and true to line and level. The various illustrations of actual formwork which follow are the best guides to good practice.

TIME FOR AND METHOD OF REMOVING FORMS.—No exact time schedule for removing forms is wise in concrete work. Concrete which is mixed wet sets slower than dry concrete and concrete sets slower in cold weather than it does in warm weather. Again the time of removal is influenced by the risk taken by too early removal, and also by the nature of the stresses in the member to be relieved of support. In all cases the forms should be removed as soon as possible so that they can be used over again and so that the concrete may be exposed to the air to hasten hardening. The following suggestions as to time of removal are general and must be followed with judgment.

Using dry concrete in warm weather the forms for retaining walls, pedestals, isolated pillars, etc., can be removed in 12 hours; using wet or sloppy concrete the time will be increased to 24 hours. In cold weather the setting is further delayed and inspection is the only safe guide to follow. Very cold weather delays setting indefinitely. Forms for small arch work like sewers and culverts may be removed in 18 to 24 hours if dry concrete is used, and in 24 to 48 hours if wet concrete is used. The time for removing large arch centers should not be less than 14 days for spans up to 50 ft. if the arch is back-filled at once; when the center is not to be used again it is better to let it stand 28 days. For very large arches the problem becomes a special one and is considered in Chapter XVII. In building construction the following schedule is a common one. Remove the column forms in 7 days and the sides of the girder forms and the floor lagging in 14 days leaving the bottom boards of the girder forms and their supports in place for 21 days.

As an example of individual practice the following requirements of a large firm of concrete contractors are given:

Walls in mass work, 1 to 3 days, or until the concrete will bear pressure of the thumb without indentation.

Thin walls, in summer, 2 days; in cold weather, 5 days.

Slabs up to 6-ft. span, in summer, 6 days; in cold weather, 2 weeks.

Beams and girders and long span slabs, in summer, 10 days or 2 weeks; in cold weather, 3 weeks to 1 month. If shores are left without disturbing them, the time of removal of the sheeting in summer may be reduced to 1 week.

Column forms, in summer, 2 days; in cold weather, 4 days, provided girders are shored to prevent appreciable weight reaching columns.

Conduits, 2 or 3 days, provided there is not a heavy fill upon them.

Arches, of small size, 1 week; for large arches with heavy dead load, 1 month.

The method of removing forms will vary in detail with the character of the structure. With proper design and lubrication of forms they will ordinarily come away from the concrete with a moderate amount of sledge and bar work. If the work will warrant it, have a special gang under a competent foreman for removing forms. The organization of this gang and the procedure it should follow will vary with the nature of the form work, and they are considered in succeeding chapters for each kind of work.

ESTIMATING AND COST OF FORM WORK.—It is common practice to record the cost of forms in cents per cubic yard of concrete, giving separately the cost of lumber and labor. This should be done, but the process of analysis should be carried further. The records should be so kept as to show the first cost per 1,000 ft. B. M. of lumber, the number of times the lumber is used, the labor cost of framing, the labor cost of erecting and the labor cost of taking down, all expressed in M. ft. B. M. In this way only is it possible to compare the cost of forms on different kinds of concrete work, and thus only can accurate predictions be made of the cost of forms for concrete work having dimensions differing from work previously done. It is well, also, to make a note of the number of square feet of exposed concrete surface to which the forms are applied.

Some of the items mentioned demand brief explanation. Framing and erecting costs are kept separate for the reason that the framing is done only once, whereas the erecting occurs two or more times. The lumber cost, where the material is used more than once, can be computed in two ways. An example will illustrate the two modes of procedure. In one of the buildings described in Chapter XIX the lumber cost $30 per M. ft. B. M. and was used three times. As 34,000 ft. B. M. were required to encase the 200 cu. yds. of concrete in one floor, including columns, it would have required 34,000 ÷ 200 = 170 ft. B. M. of lumber at $30 per M. per cubic yard of concrete if it had been used only once. But since it was used three times we may call it 170 ft. B. M. at $10 per M. per cubic yard of concrete, or we may call it 170 ÷ 3 = 57 ft. B. M. at $30 per M. per cubic yard of concrete. The authors prefer the first method, due to the fact that it is 170 ft.

B. M. that is handled and taken down each time and it is more consistent to have the lumber cost on the same basis thus:

170 ft. B. M. of lumber at $10 per M	$1.70
170 ft. B. M. handled at $2 per M	0.34
170 ft. B. M. erected at $7 per M	1.19
Total 170 ft. B. M. per cu. yd	$3.23

Returning to our main thought, there are three ways of recording the cost of form work: (1) In cents per cubic yard of concrete; (2) in cents per square foot of concrete face to which forms are applied, and (3) in dollars per 1,000 ft. B. M. of lumber used. In all cases the cost of materials and of labor should be kept separate. It is well if it can be done to attach a sketch of the forms to the record. So much for the general method of recording costs in form work.

In estimating the probable cost of forms for any job the following method will be found reliable: Having the total cubic yards of concrete in the work and the time limit within which the work must be completed determine the number of cubic yards that must be placed per day, making liberal allowances for delays. Next estimate the number of thousands of feet board measure of forms required to encase the concrete to be placed in a day. This will give the minimum amount of lumber required, for it is seldom permissible to remove the forms until the concrete has hardened over night. Now we come to the very important and puzzling question of the time element, particularly in work where it is possible to use the same forms or the same form lumber two or more times.

It has already been pointed out that wet concrete sets more slowly than dry concrete; that all concrete sets more slowly in cold than in warm weather, and that the support of forms is necessary a longer time for pieces subject to bending stress like arches and girders. General suggestions as to specific times for removing forms have also been given. Where the specifications state the time of removal the contractor has a definite guide, but where they do not, as is most often the case, he must depend very largely on judgment and previous experience. Another matter which deserves consideration is the use of the forms as staging for runways or tracks. Such use may result in forms having to stand on work for sake of their service as trestles much longer than there is any necessity so far as supporting the concrete is concerned. A derrick or cableway may often prove cheaper than tieing up form lumber by trying to make it serve the double purpose of a trestle.

The possibilities of repeated use of forms and of unit construction of forms have already been noted. This is the next point to be considered in estimating form lumber. At the expense of a little planning movable forms can be used to materially reduce the amount of lumber required. The reader is referred particularly to the chapters on retaining wall, conduit and building work for specific data on movable form work.

Having estimated the amount of lumber required and the number of times it can be used the labor cost of framing, erecting and taking down can be figured. In ordinary retaining wall work forms will cost for framing and erection from $6 to $7 per M. ft. B. M. To tear down such forms carefully and to carry the lumber a short distance will cost some $1.50 to $2 per M. ft. B. M. We have then a cost of $7.50 to $9 per M. ft. B. M. for each time the forms are erected and torn down. Where movable panels are used and the forms not ripped apart and put together again each time there is of course only the cost of moving, which may run as low as 50 cts. per M. ft. B. M. Framing and erecting centers for piers will run about the same as for retaining wall. At this point it may be noted that in estimating the cost of forms for plain rectangular piers the following method will give very accurate results. Ascertain the surface area of the four sides of the pier. Multiply this area by 2, and the product will be the number of feet board measure of 2-in. plank required. Add 40 per cent. to this, and the total will be the number of feet board measure of 2-in. plank and of upright studs (4×6), spaced 2½ ft. centers. Sometimes 3×6-in. studs are used, and spaced 2 ft. centers, which requires practically the same percentage (40 per cent.) of timber for the studs as where 4×6-in. studs are used and spaced 2½ ft. centers. No allowance is made for timber to brace the studs, since, in pier work, it is customary to hold the forms together either with bolts or with ordinary No. 9 telegraph wire, which weighs 0.06 lb. per foot. The foregoing data can be condensed into a rule that is easily remembered:

Multiply the number of square feet surface area of the sides and ends of a concrete pier by 2.8, and the product will be the number of feet board measure required for sheet plank and studs for the forms.

If the form lumber can be used more than once, divide the number of feet board measure by the number of times that it can be used, to ascertain the amount to be charged to each pier. Forms can be erected and taken down for $8 per M. carpenters being paid $2.50 and laborers $1.50 a day. Since there are 2.8 ft. B. M. of forms per square foot of surface area of concrete to be sheeted, it costs 2¼ cts. for the labor of carpenters per square foot of surface area to be sheeted. If lumber is worth $24 per M., and is used three times, then the lumber itself also costs 2¼ cts. per sq. ft. of surface area of concrete. By dividing the total number of cubic yards of concrete into the total number of square feet of area of surface to be sheeted with forms, the area per cubic yard is obtained. Multiply this area by 4½ cts., and the product is the cost per cubic yard for material in the forms (assumed to be used three times) and the labor of erecting it and taking it down.

The cost of framing and erection of forms for building work and of centers for large arches is a special problem in each case and is considered in the chapters devoted to those classes of work.

10

CONCRETE PILE AND PIER CONSTRUCTION

Two general methods of concrete pile construction are available for engineering work. By one method a hole is formed in the ground by driving a steel shell or by other special means and this hole is filled with concrete. By the other method the pile is molded in suitable forms and after becoming hard is driven as a wood or steel pile is driven. Piles constructed by the first method may be either plain or reinforced, but piles constructed by the second method are always reinforced to strengthen them for handling and driving. Concrete piers for foundation work are simply piles of enlarged diameter.

MOLDING PILES IN PLACE.—Molding piles in place requires the use of special apparatus, and this apparatus is to a very large degree controlled by patents. Pile work of this kind is thus generally done by concerns which control the use of the apparatus employed and the general contractor can undertake it only by permission of the proprietary companies. The methods of work followed and the cost of work are thus of direct interest only as general information.

Method and Cost of Constructing Raymond Piles.—The machinery and processes employed in the construction of Raymond concrete piles are patented and all piling work by this method is controlled by the Raymond Concrete Pile Co. As detail costs of construction are not given out by the company the following figures collected by the authors are subject to revision. They are believed to be fairly approximate, having in one case been obtained by personal watch on the work and in the other case from authentic records of the engineers on the work.

The pile is made as follows: A collapsible steel core 30 ft. long, 20 ins. diameter at the top and 6 ins. diameter at the bottom, encased in a thin sheet steel shell, is driven into the ground by an ordinary pile driver. When it has reached the proper depth, a wedge is loosened, permitting the two sections of the core to come closer together so that the core can be pulled out of the hole, leaving the steel shell behind as a casing to prevent the sides from caving in.

The shell is made of No. 20 gage steel, usually in four or more sections, which telescope one over the other. A nest of sections is slipped over the lower end of the core as it hangs in the leads, a rope is hitched around the outer section and the engine hoists away until the sections are "un-telescoped" and drawn snug onto the core. The rope is then unfastened and the driving begins. Figure 49 shows the usual pile driving rig used. The following are examples of pile construction in actual work:

Example I.—In this work, for a building foundation in New York City, the pile driver was mounted on a turntable, the framework of the turntable in turn resting on rollers traveling on timbers laid on the ground. The driver was moved along and rotated when necessary by ropes passing around the winch head of the engine. The driver had 50-ft. leads and a 3,100-lb. hammer operated by an ordinary friction clutch hoisting engine. The hammer blow was received by an oak block fitting into a recess at the top of the steel core.

This block was so battered by the blows that it had to be renewed about every five or six piles driven. A ¾-in. wire rope passing over a 10-in. sheave lasted for the driving of 130 piles and then broke. When the work was first begun the crew averaged 10 piles per 10-hour day, but the average for the job was 13 piles per day, and the best day's work was 17 piles. The cost of labor and fuel per pile was as follows:

Fig. 49.—Pile Driver Rigged for Constructing Raymond Concrete Piles.

5 men on driver at $1.75	$ 8.75
2 men handling shells at $1.75	3.50
1 engineman	3.00
6 men mixing and placing concrete	10.50
1 foreman	5.00
Coal and oil	2.50

Total, 13 piles, at \$2.55	\$33.25

Deducting the cost of placing the concrete we get a cost of \$1.75 for driving the cores. The pile, 25 ft. long, 6 ins. at the point and 18 ins. at the head, contains 21¼ cu. ft., or 0.8 cu. yd., of concrete, and has a surface area of 77 ft. As No. 20 steel weighs 1.3 lbs. per sq. ft., each shell weighed approximately 100 lbs. The cost per pile may then be summarized as follows:

1.2 bbls. cement in 0.8 cu. yd., at \$1.75	\$2.10
0.8 cu. yd. stone at \$1.25	1.00
⅓ cu. yd. sand at \$1.05	0.35
100 lbs. steel in shell at 3½ cts.	3.50
Labor and fuel as above	2.55
Total per pile (38 cts. per lin. ft.)	\$9.50

This cost, it should be carefully noted, does not include cost of moving plant to and from work or general expenses.

Example II.—In constructing a building at Salem, Mass., 172 foundation piles, 14 to 37 ft. long, 6 ins. diameter at the point and 20 ins. diameter at the top, were constructed by the Raymond process. The general contractor made the necessary excavations and provided clear and level space for the pile driver, braced all trenches and pier holes, set stakes for the piles and gave all lines and levels. The piles were driven by a No. 2 Vulcan steam hammer with a 3,000-lb. plunger having a drop of 3 ft., delivering 60 blows per minute. Figure 49 shows the driver at work. Sixteen working days were occupied in driving the piles after the driver was in position. The greatest number driven in one day was 20, and the average was 11 piles per day When in position for driving, the average time required to complete driving was 12 minutes. The total number of blows varied from about 310 to 360, the average being about 350. The piles were driven until the penetration produced by 8 to 10 blows equaled 1 in. When in full operation, a crew of 5 men operated the pile driver. Seven men were engaged in making the concrete and 5 men working upon the metal shells.

Assuming the ordinary organization and the wages given below, we have the following labor cost per day:

1 foreman at $5	$ 5.00
1 engineman at $3	3.00
4 laborers on driver at $1.75	7.00
6 laborers making concrete at $1.75	10.50
5 laborers handling shells at $1.75	8.75
Total	$34.25

As 172 piles averaging 20 ft. in length were driven in 16 days, the total labor cost of driving, given by the figures above, is 16 × $34.25 = $548, or practically 16 cts. per lineal foot of pile driven.

The concrete used in the piles was a 1-3-5 Portland cement, sand and 1½-in. broken stone mixture. A 20-ft. pile of the section described above contains about 20 cu. ft. of concrete, or say 0.75 cu. yd. We can then figure the cost of concrete materials per pile as follows:

0.85 bbl. cement at $1.60	$1.36
0.36 cu. yd. sand at $1	0.36
0.60 cu. yd. stone at $1.25	0.75
Total per pile	$2.47

The steel shell has an area of about 72 sq. ft., and as No. 20 gage steel weighs 1.3 lbs. per sq. ft., its weight for each pile was about 94 lbs. Assuming the cost of coal, oil, etc., at $2.50 per day, we have the following summary of costs:

	Per lin. ft. of pile.
Labor driving and concreting	$0.16
Concrete materials	0.123
94 lbs. steel shell at 3 cts.	0.145
Coal, oil, etc.	0.011

Total	$0.439

The cost does not include interest on plant, cost of moving plant to and from work and general expenses.

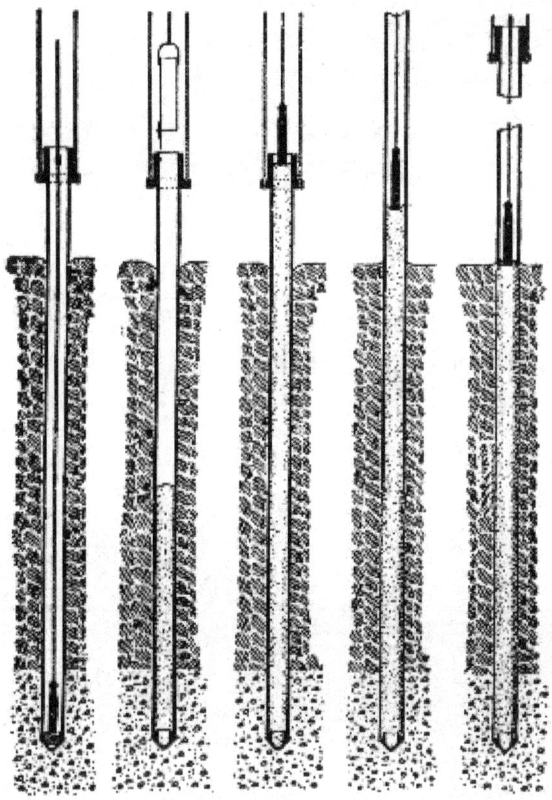

Fig. 50.—Sketch Showing Method of Constructing Simplex Concrete Piles.

Method of Constructing Simplex Piles.—The apparatus employed in driving Simplex piles resembles closely the ordinary wooden pile driven, but it is much heavier and is equipped to pull as well as to drive. A 3,300-lb. hammer is used and it strikes on a hickory block set in a steel drive head which rests on the driving form or shell. This form consists of a ¾-in. steel shell 16 ins. in diameter made in a single 40-ft. length. Around the top of the shell a ½-in. thick collar or band 18 ins. deep is riveted by 24 1-in. countersunk rivets. This band serves the double purpose of preventing the shell being upset by the blows of the hammer and of giving a grip for fastening the pulling tackle. The bottom of the form or shell is provided with a point. Two styles of point are employed. One style consists of two segments of a cylinder of the same size as the form, so cut that they close together to form a sort of clam shell point. In driving, the two jaws are held closed by the pressure of the earth and in pulling they open apart of their own weight to permit the concrete to pass them. This point, known as the alligator point, is pulled with the shell. It is suitable only for driving in firm, compact soil, in loose soil the pressure inward of the walls keeps the jaws partly closed and so contracts the diameter of the finished pile. The second style of point is a hollow cast iron point, 10 ins. deep and 16½ ins. in diameter, having a neck over which the driving form slips and an annular shoulder outside the neck to receive the circular edge of the shell. The projected sectional area of this point is 1.4 sq. ft. It is left in the ground when the form is withdrawn. The form is withdrawn by means of two 1-in. cables fastened to a steel collar which engages under the band at the top of the form. The cables pass in the channel leads on each side over the head of the driver and down in back to a pair of fivefold steel blocks, the lead line from which passes to one of the drums of the engine. In this manner the power of the drum is increased ten times and it is not unusual to break the pulling cables when the forms are in hard ground. The general method of construction is about as shown by Fig. 50, being changed slightly to meet varying conditions. The form resting on a cast iron point is driven to hard ground. A heavy weight is then lowered into the form to make sure the point is loose. While the weight is at the bottom of the form a target is placed on its line at the top of the form, the purpose of which will be apparent later. The weight is then withdrawn. Given the length of the pile and sectional area, it is an easy matter to determine the volume of concrete necessary to fill the hole.

This amount is put into the form by means of a specially designed bottom dump bucket, which permits the concrete to leave it in one mass, reaching its destination with practically no disintegration. It will be noticed that when the full amount of concrete is in the form its surface is considerably above the surface of the ground. This is due to the fact that the thickness of the form occupies considerable space that is to be occupied by the concrete. The weight is now placed on top of the concrete and the form is pulled. The target previously mentioned now becomes useful. As the form is withdrawn the concrete settles down to occupy the space left by the walls of the form. Obviously this settlement should proceed at a uniform rate, and as it is difficult to watch the weight, the target on its line further up is of considerable help. By watching this target in connection with a scale on the leads of the driver, it can be readily told how the concrete in the form is acting. As another check, the target, just as the bottom of the form is leaving the ground should be level with the top of the form.

This would indicate that the necessary amount of concrete has gone into the ground and that, other conditions being all right, the pile is a good one. In some grounds where the head of concrete in the form exerts a greater pressure than the back pressure or resistance of the earth, the concrete will be forced out into the sides of the hole, making the pile of increased diameter at that point and necessitating the use of more concrete to bring the pile up to the required level.

Method of Constructing Piles with Enlarged Footings.—A pile with an enlarged base or footing has been used in several places by Mr. Charles R. Gow of Boston, Mass., who has patented the construction. A single pipe or a succession of pipes connected as the work proceeds is driven by hammer to the depths required. The material inside the shell is then washed out by a water jet to the bottom of the shell and then for a further distance below the shell bottom.

An expanding cutter is then lowered to the bottom of the hole and rotated horizontally so as to excavate a conical chamber, the water jet washing the earth out as fast as it is cut away. When the chamber has been excavated the water is pumped out and the chamber and shell are filled with concrete. The drawings of Fig. 51 show the method of construction clearly.

The chambering machine is used only in clay or other soil which does not wash readily. In soil which is readily washed the chamber can be formed by the jet alone. The practicability of this method of construction is stated by Mr. Gow to be limited to pipe sizes up to about 14 ins. in diameter.

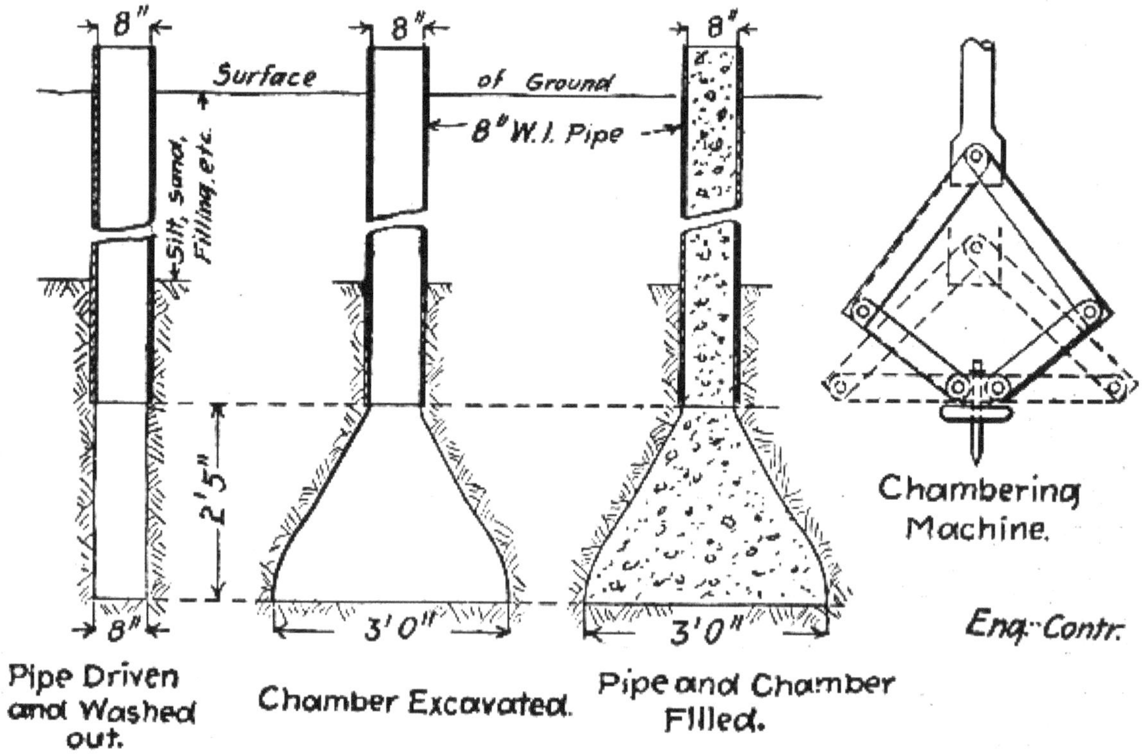

Fig. 51.—Sketch Showing Method of Constructing Concrete Piles with Enlarged Footings.

Method of Constructing Piles by the "Compressol" System.—The compressol system of concrete pile or pillar construction is a French invention that has been widely used abroad and which is controlled in this country by the Hennebique Construction Co., of New York, N. Y. The piles are constructed by first ramming a hole in the ground by repeatedly dropping a conical "perforator" weighing some two tons. This perforator is raised and dropped by a machine resembling an ordinary pile driver. The conical weight gradually sinks the hole deeper and deeper by compacting the earth laterally; this lateral compression is depended upon so to consolidate the walls of the hole that they do not cave before the concrete can be placed. The concrete is deposited loose in the hole and rammed solid by dropping a pear-shaped weight onto it as it is placed. The view Fig. 52 shows the "perforator" and the tamping apparatus at work. Very successful work has been done abroad by this method.

Fig. 52.—View of Apparatus Used in Constructing Compressol Piles.

Method of Constructing Piers in Caissons.—For piles or pillars of diameters larger than say 18 ins. the use of driving shells and cores becomes increasingly impracticable. Concrete pillars of large size are then used. They are constructed by excavating and curbing a well or shaft and filling it with concrete. This construction has been most used in Chicago, Ill., for the foundations for heavy buildings, but it is of general application where the sub-soil conditions are suitable. The method is not patented or controlled by patents in any particular, except that certain tools and devices which may be used are proprietary.

General Description.—The caisson method of construction is simple in principle. A well is dug by successive excavations of about 5 ft. each. After each excavation of 5 ft. is completed, wood lagging is placed around the sides and supported by internal steel rings, so that the soft ground around the excavation is maintained in its former position. The methods of excavating and removing the soil and of constructing the lagging are considered in detail further on. The caissons vary in diameter according to the load; some as large as 12 ft. in diameter have been sunk, but the usual diameter is 6 ft.; a caisson of 3 ft. in diameter is as small as a man can get into and work. When the pier goes to bed rock the caisson is made of uniform diameter from top to bottom, but where the pier rests on hardpan the bottom portion of the well is belled out to give greater bearing area. It is customary to load the piers about 20 tons per square foot.

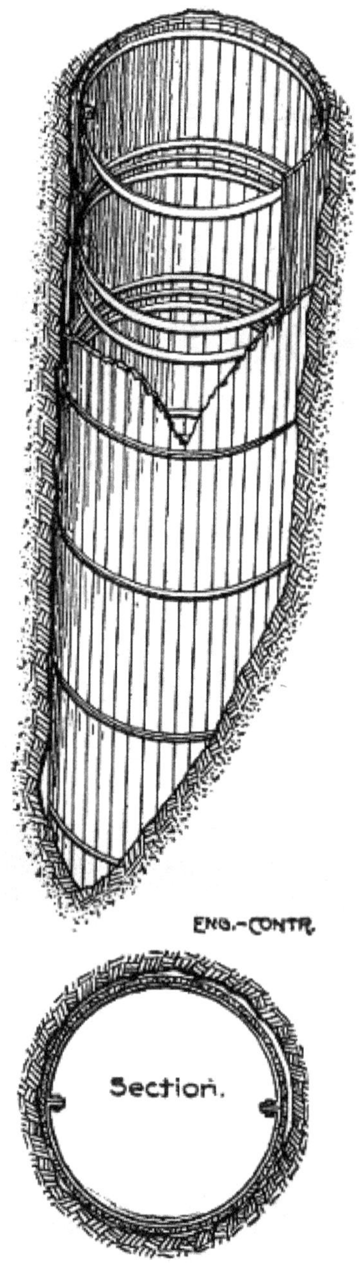

Fig. 53.—Curbing for Concrete Piers (Usual Construction).

Caisson Construction.—The caisson construction, or more correctly the form of curbing most commonly used, is that indicated by the sketch, Fig. 53. The lagging is 2×6 in. or 3×6 in., stuff 5 ft. 4 ins. or 4 ft. long set vertically around the well and held in place by interior wrought iron rings. For a 6-ft. diameter caisson these hoops are ¾ by 3 ins.; they are made in two parts, which are bolted together as shown by Fig. 53. Generally there are two rings for each length of lagging; for 5-ft. lagging they are placed about 9 ins. from each end. In some cases, however, engineers have specified three rings for the upper sections in soft clay and two rings for the sections in the hard ground lower down. The lagging used is not cut with radial edges, but is rough, square cut stuff; the rings, therefore, take the inward pressure altogether.

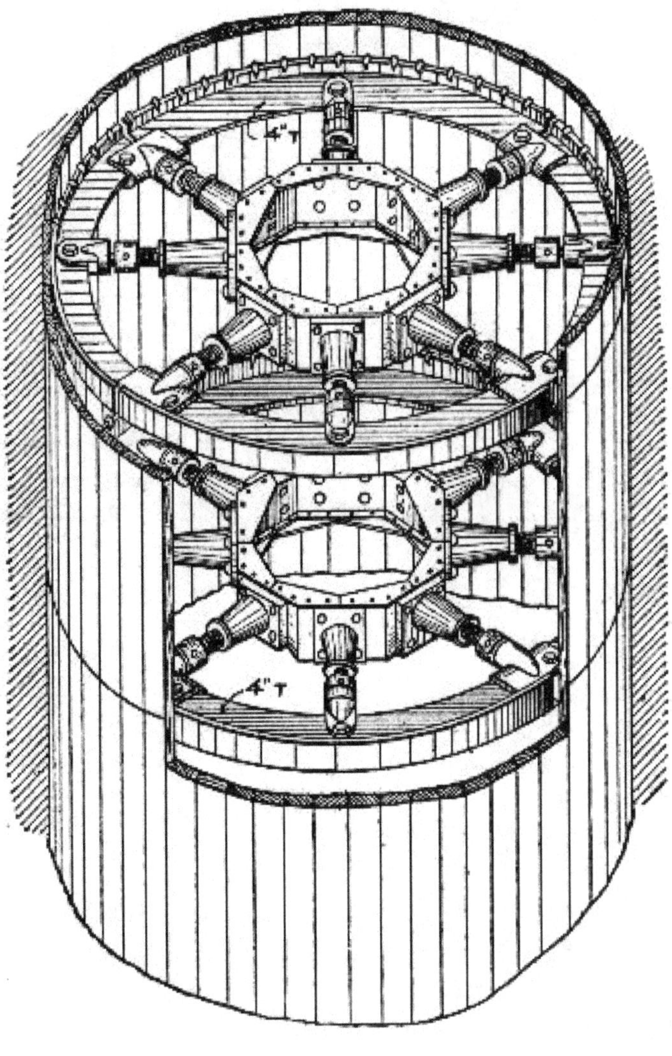

Fig. 54.—Curbing for Concrete Piers (Jackson Patent).

In some recent work done by the inventor use has been made of the caisson construction shown by Fig. 54 and patented by Mr. Geo. W. Jackson. In place of the plain rings a combination of T-beam ribs and jacks is used; this construction is clearly shown by the drawing. The advantages claimed for the construction are that it gives absolute security to the workmen and the work, that the lagging can be jacked tightly against the outer walls of the well, that the braces form a ladder by which the workmen can enter and leave the well, and that the possibility of shifting the bracing easily permits the concrete to be placed to the best advantage. On the other hand the braces abstruct the clear working space of the caissons.

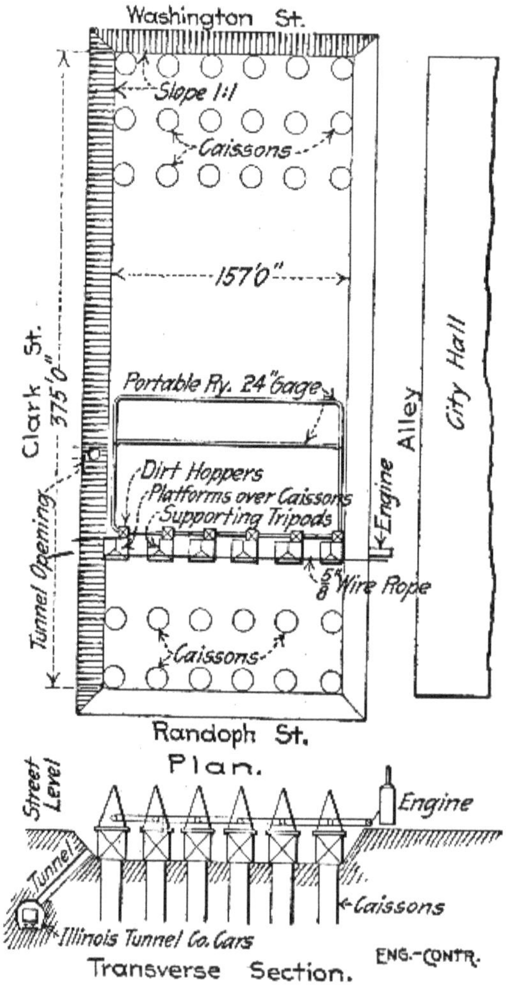

Fig. 55.—Layout of Plant for Concrete Pier Construction. Cook County Court House Foundations.

Excavating and Handling Material.—The excavation of the wells is done by hand, using shovels and picks, and, in the hardpan, special grubs made by A. J. Pement and George Racky, Chicago blacksmiths. The excavated material is hoisted out of the well in buckets made by the Variety Iron Works, of Chicago. For caissons which are not specified to go to rock it is considered more economical to do the hoisting by windlass derricks operated by hand. These derricks have four 6×6-in. legs and a 3×6-in. top piece. When the caissons go to rock the hoisting is done by power, so-called "cable set-ups" being used in most cases. To illustrate this method the following account of the foundation work for the Cook County Court House is given:

The Cook County Court House foundations consist of 126 caissons varying from 4 ft. to 10½ ft. in diameter and averaging$ 7½ ft. in diameter. They were sunk to rock at a depth of 115 ft. below street level. The work involved 22,000 cu. yds. of excavation and the placing in the caissons of 17,000 cu. yds. of concrete. Over 1,000 piles about 40 ft. long, that had formed the foundation of the old Court House built in 1875, were removed. These piles were found to be in good condition. The work was done by the George A. Fuller Co., of Chicago, Ill., Contractors, with Mr. Edgar S. Belden Superintendent in Charge. The details which follow have been obtained from Mr. Belden.

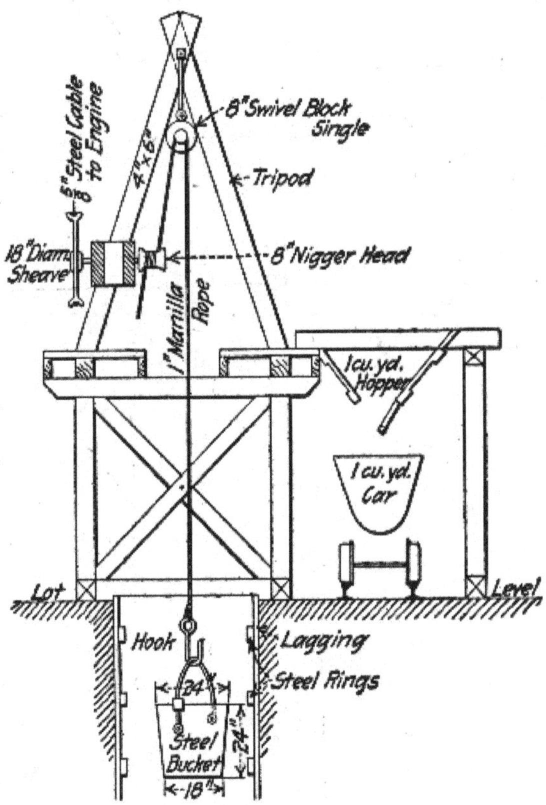

Fig. 56.—Section Showing Arrangement of Hoist for Concrete Pier Construction.

The foundation area was 157×375 ft., and was excavated to a depth of 15 ft. below the street surface before the caissons were started. The caissons, of which there were 126, were arranged in rows across the lot, there being from six to eight caissons in a row. The arrangement of the plant for the work is indicated by Fig. 55. One row of caissons formed a unit. A platform or "stand" was erected over each caisson and carried in its top a tripod fitted with a "nigger head" operated by a rope sheave. This arrangement is shown by Fig. 56. An engine on the bank operated by a rope drive all the tripod sheaves for a row of six or eight caissons. The arrangement is indicated by Fig. 55. The clay hoisted from the pits was dumped into 1 cu. yd. hoppers with which the stands were fitted, as shown by Fig. 56; when a hopper was full it was dumped into a car running on a 24-in. gage portable track. Side dump Koppel cars of 1 cu. yd. capacity were used; they dumped their load into an opening connected with the tracks of the Illinois Tunnel Co., where the material passed into tunnel cars and was taken to the lake front about one mile away. As soon as one row of caissons was completed the stands, tripods, etc., which were made portable, were shifted to another row. At times as many as five units were in operation, sinking 40 caissons.

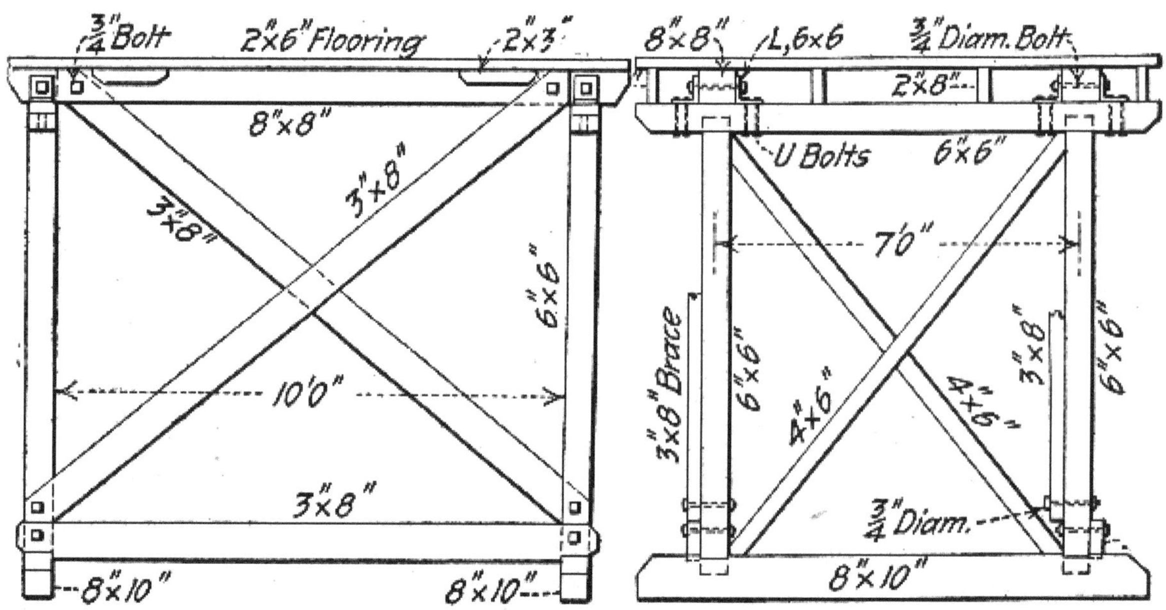

Side Elevation. End Elevation.

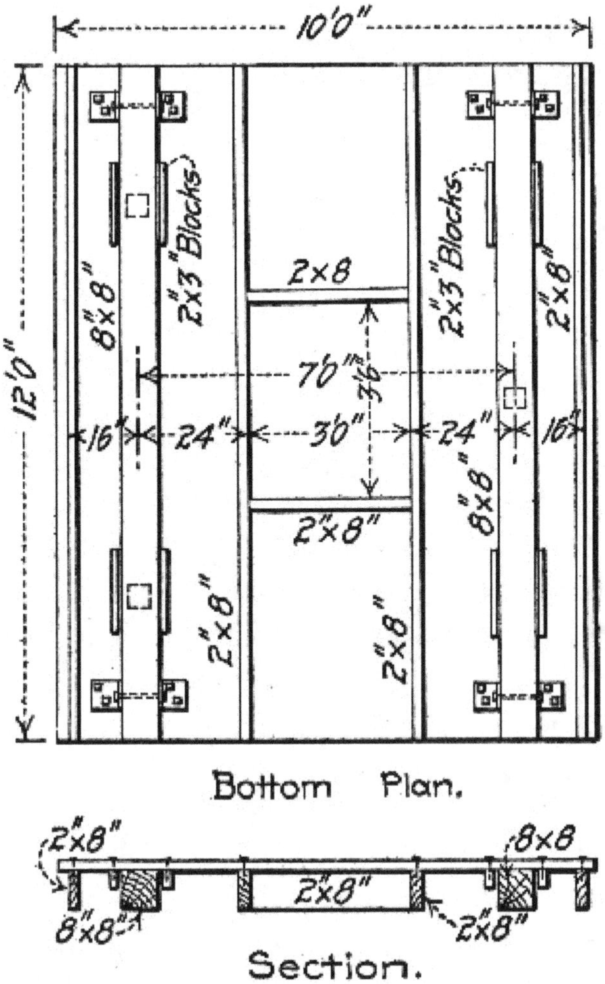

Fig. 57.—Details of Working Platform for Concrete Pier Construction.

Fig. 56 shows the arrangement in detail at one caisson. In this work the lagging used was 3×6-in. maple, 5 ft. 4 ins. long, and was supported by 3×¾-in. steel hoops. The lagging was matched and dressed. The "nigger head," as will be seen, is operated by a rope sheave on the same axle. As stated above, an endless rope drive operated all the "nigger heads" on a row of caissons. A 26-in. driving sheave was attached to an ordinary hoisting engine equipped with a governor. The driving rope was ⅝-in. steel. It was wrapped twice around the driving sheave and once around the "nigger head" sheaves. These latter were 18 ins. in diameter. For the hoists 1-in. Manila rope was used. The other details, the bucket, bucket hook, swivel block, etc., are made clear by the drawing. The platforms, tripods, etc., were of the standard dimensions and construction adopted by the contractors of the work. Detail drawings of the standard platform are given by Fig. 57. One of these platforms contains about 1,000 ft. B. M. of lumber. As will be seen, all connections are bolted, no nails being used anywhere. A platform can thus be taken down and stored or shipped and erected again on another job with very little trouble.

The plant described handled some 22,000 cu. yds. of excavated material on this work. Work was kept up night and day, working three 8-hour shifts. It took an average of 35 shifts to excavate one row of caissons. No figures of the working force or the cost of excavation of this work are available.

Mixing and Placing Concrete.—The placing of the concrete in the excavated wells is done by means of tremies, or, which is more usual, by simply dumping it in from the top, workmen going down to distribute it. The manner of mixing the concrete and of handling it to the caisson varies of course with almost every job. As an example of the better arranged mixing and handling plants the one used on the Cook County Court House work may be described. This plant is shown by the sketch, Fig. 58.

Bins for the sand and stone were built at one side of the lot on the sloping bank; their tops were level with the street surface and their bottoms were just high enough to permit their contents to be delivered by chutes into 1 cu. yd. cars. Wagons dumping through traps in the platform over the bin delivered the sand and stone. The sketches indicate the arrangement of the bins and mixer and the car tracks connecting them. The raw material cars were first run under the stone bin and charged with the required proportion of stone, and then to the sand bin, where the required proportion of sand was chuted on top of the stone. The loaded car was then hauled up the incline and dumped into the hopper, where cement and water were added. A No. 2½ Smith mixer was used and discharged into cars which delivered their loads on tracks leading to the caissons. The same cars and portable tracks were used as had been used to handle the excavated material. In operation a batch of raw materials was being prepared in the hopper while the previous batch was being mixed and while the concrete car was delivering the still previous batch to the caissons. An average of 40 batches an hour mixed and put into the caissons was maintained with a force of 25 men. In all some 17,000 cu. yds. of concrete were mixed and deposited.

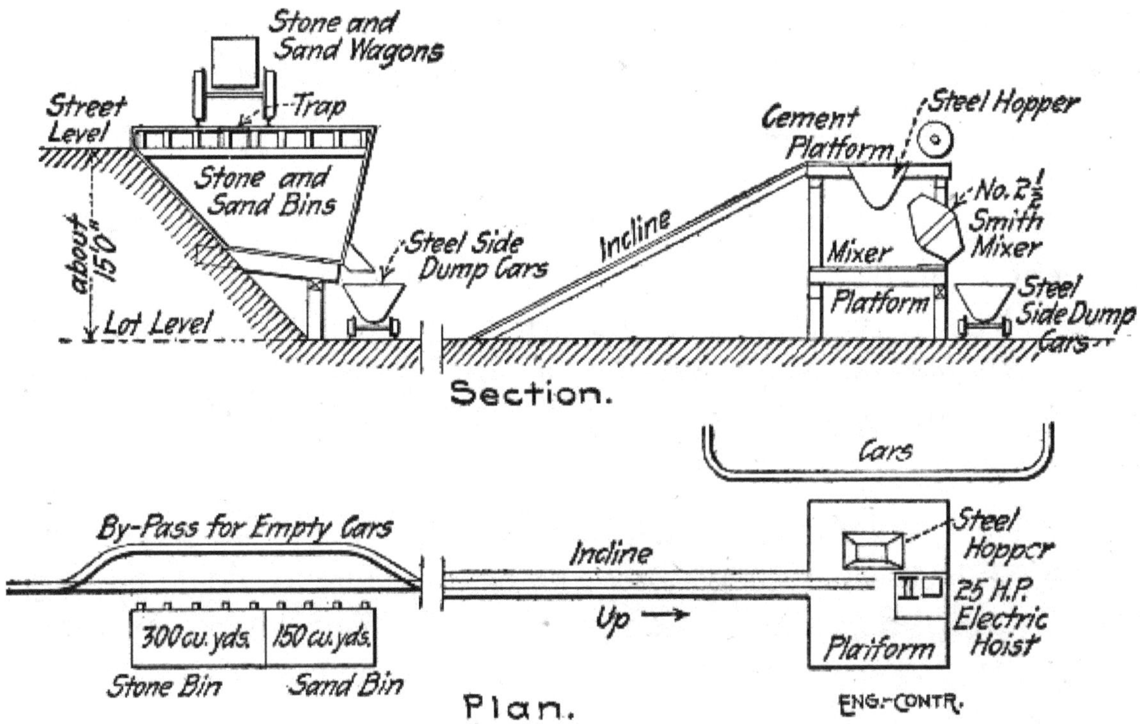

Fig. 58.—Arrangement of Concrete Making Plant, Concrete Pier Construction.

Cost of Caisson Work.—The following attempt to get at the cost of caisson work is based largely upon information obtained from Mr. John M. Ewen, John M. Ewen Co., Engineers and Builders, Chicago, Ill. Mr. Ewen says:

"My experience has taught me that it is almost impossible to determine any definite data of cost for this work. This is due to the fact that no two caisson jobs will average the same cost, notwithstanding the fact that the cost of material used and the labor conditions are exactly the same. This condition is due to the great variety in texture of the soil gone through. For instance, it has come under my experience that in caissons of the same diameter on the same job it required but fifteen 8-hour shifts to reach bedrock in some of these, while it required as many as 21 to 25 shifts to reach rock in the others, rock being at the same elevation. In fact, the digging all the way to rock in some was the best that could be wished for, while in the others boulders and quicksand were encountered, and the progress was slower, and the cost consequently greater.

"Again, we have known it to require eight hours for two men to dig 8 ins. in hardpan in one caisson, while on a job going on at the same time and on the opposite corner of the street two men made progress of 2 ft. in 8 hours through apparently the same stuff, the depth of hardpan from grade being 61 ft. 6 ins. in both instances, and the quality of labor exactly the same.

"There have been more heavy losses among contractors due to the unexpected conditions arising in caisson digging than in any other item of their work, and I predict a loss to some of them that will be serious indeed if an attempt be made to base future bids for caisson work entirely upon the data kept by them on past work. If a contractor is fortunate enough to find the ordinary conditions existing in his caisson work, and by ordinary conditions I mean few boulders, no quicksand, ordinary hardpan and no gas, the following items may be considered safe for figuring caisson work:

"Figure that it will require from 22 to 25 shifts of 8 hours each to strike bedrock, bedrock being from 90 to 95 ft. below datum, and datum being 15 ft. below street grade; figure 2 diggers to the shift in all caissons over 5 ft. in diameter, 45 cts. per hour for each digger; figure 1 top man at 40 cts. per hour, and 1 mucker or common laborer at 30 cts. per hour for all caissons in which there are two diggers, and 1 top man less if 1 digger is in the caisson, which condition exists generally in caissons less than 5 ft. in diameter. Add the cost of ⅝-in. cable, tripods, sheaves, 1-in. Hauser laid line, nigger heads, ball-bearing blocks, etc., for rigging of the job. Lagging, which is 2×6 ins. and 3×6 ins. hemlock or some hard wood, in length of 5 ft. 4 ins. and 4 ft., is priced all the way from $20 to $22.50 and $21 to $24.50 per M. ft. B. M., respectively. The price of caisson rings is $2.40 per 100 lbs. The cost of specially made grubs for digging in hardpan is about $26 per dozen. Shovels are furnished by the diggers themselves in Chicago, Ill. The cost of temporary electric light is $10 per caisson. This includes cost of cable, lamps, guards, etc. Add the cost of or rental of engine or motors for power.

"Some engineers specify three rings to be used to each set of lagging below the top set until hardpan is reached, then two rings for each of the remaining sets from hardpan to rock. This is, of course, to insure against disaster from great pressure of the swelling clay above the hardpan strata, and may or may not be necessary. These rings are ¾×3 ins. wrought iron.

"For caissons which are not specified to go to rock, it is not considered economical to rig up cable set-ups, but rather to use windlass derricks. In this case 1-in. Hauser laid line is used as the means of hoisting the buckets of clay out of the caisson, as is the case in cable set-ups, hand power being used on the windlass derricks instead of steam or electricity. The windlass derricks are made with four legs out of 6×6-in. yellow pine lumber. The top piece is generally a piece of 3×6-in. lagging. The cost of windlass and boxes is about $35 per dozen. Hooks for caisson buckets cost 45 cts. each. Caisson buckets cost $8 each.

"With the above approximate units as a basis, I have seen unit prices given per lineal foot in caisson work which ranged all the way from $12 to $16.50 for 6-ft. diameter caissons, larger and smaller sized caissons being graded in price according to their size. This unit price included rings, lagging, concrete, power, light, labor, etc."

From the above data the following figures of cost can be arrived at, assuming a 6-ft. caisson:

Labor.	Per day.
2 diggers in caisson, at $3.60	$ 7.20
1 top man, at $3.20	3.20
1 mucker, at $2.40	2.40
	$12.80

The depth sunk varies from 3½ to 8 ft. per 8-hour day, depending on the material. Assuming an average of 4 ft., we have then 4 lin. ft. of caisson, or 2.8 cu. yds. excavated at a labor cost of $12.80, which is at the rate of $3.20 per lin. ft., or $4.57 per cu. yd. We now get the following:

	Per lin. ft. Caisson.
40 ft. B. M. (2×6-in. lagging) at $25	$1.00
60 lbs. iron (¾×3-in. rings) at 2½ c.	1.50
0.7 cu. yd. excavation at $4.57	3.20
0.7 cu. yd. muck hauled away at $1	0.70
0.7 cu. yd. concrete at $5	3.50
Electric light	0.10
Total	$10.00

If 3×6-in. lagging is used add 50 cts. per lin. ft. of caisson.

MOLDING PILES FOR DRIVING.—Piles for driving are molded like columns in vertical forms or like beams in horizontal forms. European constructors have a strong preference for vertical molding, believing that a pile better able to withstand the strain of driving is so produced; such lamination as results from tamping and settling is, in vertical molding, in planes normal to the axis of the pile and the line of driving stress. Vertical molding has been rarely employed in America and then only for molding round piles. The common belief is that horizontal molding is the cheaper method. In the ordinary run of work, where comparatively few piles are to be made, it is probably cheaper to use horizontal molds, but where a large number of piles is to be made, the vertical method has certain economic advantages which are worth considering.

Fig. 59.—Plant for Vertical Molding of Concrete Piles.

Vertical molding necessitates a tower or staging to support the forms and for handling and placing the concrete; an example of such a staging is shown by Fig. 59. To counterbalance this staging, horizontal molding necessitates a molding platform of very solid and rigid construction if it is to endure continued and repeated use. In the matter of space occupied by molding plant, vertical molding has the advantage. A tower 40 ft. square will give ample space around its sides for 80 vertical forms for 12-in. piles and leaves 1 ft. of clear working space between each pair of forms. The ground area occupied by this tower and the forms is 1,764 sq. ft. With the same spacing of molds a horizontal platform at least 25 × 160 ft. = 4,000 sq. ft., would be required for the molds for the same number of piles 25 ft. long. For round piles, vertical molding permits the use of sectional steel forms; horizontal forms for round piles are difficult to manage. For square piles vertical molding requires forms with four sides; horizontal forms for square piles consist of two side pieces only, the molding platform serving as the bottom and no top form being necessary. Thus, for square piles horizontal molding reduces the quantity of lumber per form by 50 per cent. The side forms for piles molded on their sides can be removed much sooner than can the forms for piles molded on end, so that the form material is more often released for reuse. The labor of assembling and removing forms is somewhat less in horizontal molding than in vertical molding. Removing the piles from molding bed to storage yard for curing requires derricks or locomotive cranes in either case and as a rule this operation will be about as expensive in plant and labor in one case as in the other. In the ease and certainty of work in placing the reinforcement, horizontal molding presents certain advantages, the placing and working of the concrete around the reinforcement is also easier in horizontal molding. Mixing and transporting the concrete materials and the concrete is quite as cheap in vertical molding as in horizontal molding. If anything, it is cheaper with vertical molding, since the mixer and material bins can be placed within the tower or close to one side where a tower derrick can hoist and deposit the concrete directly into the molds. Car tracks, cars, runways and wheelbarrows are thus done away with in handling the concrete from mixer to molds. Altogether, therefore, the choice of the method of molding is not to be decided off-hand.

DRIVING MOLDED PILES.—Driving molded concrete piles with hammer drivers is an uncertain operation. It has been done successfully even in quite hard soils and it can be done if time is taken and the proper care is exercised. The conditions of successful hammer driving are: Perfect alignment of the pile with the line of stroke of the hammer; the use of a cushion cap to prevent shattering of the pile-head, and a heavy hammer with a short drop. The pile itself must have become well cured and hardened. At best, hammer driving is uncertain, however; shattered piles have frequently to be withdrawn and the builder is never sure that fractures do not exist in the portion of the pile that is underground and hidden. The actual records of concrete pile work given in succeeding sections illustrate successful examples of hammer driving. The plant required need not vary from that ordinarily used for driving wooden piles, except that more power must be provided for handling the heavier concrete pile and that means must be provided for holding the pile in line and protecting its head.

Sinking concrete piles by means of water jets is in all respect a process similar to that of jetting wooden piles. Examples of jetting are given in succeeding section. In rare cases, driving shells, or sheaths have been used for driving molded piles.

Method and Cost of Molding and Jetting Piles for an Ocean Pier.—In reconstructing in reinforced concrete the old steel pier at Atlantic City, N. J., some 116 reinforced concrete piles 12 ins. in diameter were molded in air and sunk by jetting. The piles varied in length with the depth of the water, the longest being 34½ ft. Their construction is shown by Fig. 60, which also shows the floor girders carried by each pair of piles and forming with them a bent, and the struts bracing the bents together. In molding and driving the piles the old steel pier was used as a working platform.

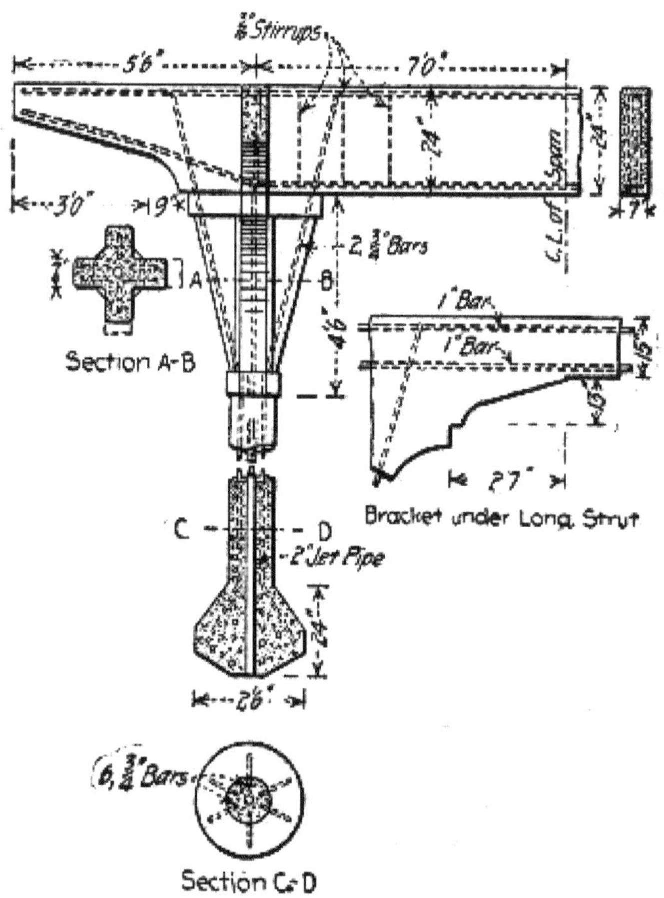

Fig. 60.—Concrete Pile for Pier at Atlantic City, N. J.

The forms for the piles were set on end on small pile platforms located close to the positions to be occupied by the piles and were braced to the old pier. The forms were of wood and the bulb point, the shaft and the knee braces were molded in one piece. Round iron rods were used for reinforcement. The concrete was composed of 1 part Vulcanite Portland cement, 2 parts of fine and coarse sand mixed and 4 parts of gravel 1 in. and under in size. The mixture was made wet and was puddled into the forms with bamboo fishing rods, which proved very efficient in working the mixture around the reinforcing rods and in getting a good mortar surface. The concrete was placed in small quantities; it was mostly all hand mixed. The forms were removed in from 5 to 7 days, depending on the weather.

The piles were planned to be sunk by water jet and to this end had molded in them a 2-in. jet pipe as shown. They were sunk to depths of from 8 ft. to 14 ft. into the beach sand. Water from the city water mains at a pressure of 65 lbs. per sq. in. was used for jetting; this water was furnished under special ordinance at a price of $1 per pile, and a record of the amount used per pile was not kept. The piles were swung from the molding platforms and set by derricks and block and fall. The progress of jetting varied greatly owing to obstructions in places in the shape of logs, old iron pipes, etc. In some cases several days were required to get rid of a single pipe. In clear sand, with no obstruction, a 12-in. pile could be jetted down at the rate of about 8 ft. per hour, working 1 foreman and 6 men. The following is the itemized actual cost of molding and sinking a 26-ft. pile with bulb point and knee braces complete:

Forms—		Cost per pile.
Lumber, 340 ft. B. M. @ $30	$10.20	...
Labor (carpenters @ $2.50 per day)	12.00	...
Oil, nails, oakum, bolts, clamps, etc.	1.20	...
	$23.40	$ 3.90
Times used	6	
Reinforcement—		
275 lbs. of plain ¾-in. steel rods @ 2 cts. per lb.	$ 5.50	...

Preparing and setting, 4/10 ct. per lb.	1.10	6.60
Jet Pipe—		
26½ ft. of 2-in. pipe @ 10 cts. per ft. in place.	2.65	...
Setting Forms—		
6 men @ $2.50 per day = $15, set 4 piles	3.75	...
Material—		
90/100 Cu. yds. gravel @ $1.50 per yd.	1.35	...
45/100 cu. yds. sand @ $1.50 per yd.	.67	...
1.50 bbls. cement @ $1.60	2.40	4.42
Labor—		
Concrete and labor foreman	3.00	...
6 laborers, mixing and placing by hand, $1.75 each	10.50	...
	$13.50	3.38
Average number of piles concreted per day	4	
Removing Forms—		
4 men @ $2.50 remove and clean in half day 4 columns	1.25	...
1 man @ $2.25 plastering column with cement grout (4 per day)	.56	...
Jetting 10 ft. into Sand—		
Foreman	$ 3.00	...
4 men, $2.25 each, handling hose and traveler	9.00	...
	$12.00	3.00
Average number of piles jetted per day	4	...
City water pressure used for jetting @ $1 per pile	...	1.00
Superintendence @ $5.00 per day	...	1.25
Caring for trestle, traveler, material, etc.	...	4.84
Total cost per pile		$36.60

The pile being 26 ft. long, the cost in place was $1.41 per foot. Subtracting the cost of sinking amounting to $7.09 per pile, we have the cost of a 26-ft.

pile molded and ready to sink coming to about $1.10 per foot. It should be noted that this is the cost for a pile of rather complicated construction; a plain cylindrical pile should be less expensive.

Method of Molding and Jetting Square Piles for a Building Foundation.—The foundation covered about an acre. The soil was a deposit of semi-fluid mud and quicksand overlying a very irregular rock bottom and encircled by a ledge of rock. The maximum depth of the mud pocket was 40 ft., and interspersed were floating masses of hard pan. Soundings were made at the locations of all piles; a ½-in. gas pipe was coupled to a hose fed by city pressure and jetted down to rock, the depth was measured, the sounding was numbered and the pile was molded to length and numbered like the sounding. In all 414 piles were required, ranging in length from 1½ to 40 ft.; all piles up to 6 ft. were built in place in wooden forms. The piles were 13 ins. square and were of 1-2½-4 concrete reinforced with welded wire fabric. A tin speaking tube was molded into each pile at the center. This tube was stopped about 10 ins. from the head and by means of an elbow and threaded nipple projected through the side of the pile to allow of attaching a pressure hose. The piles were handled to the pile driver, the hose attached and water supplied at 100 lbs. pressure by a pump. Churning the pile up and down aided the driving. A hammer was used to force the piles through the hard pan layers. A wooden follower was used to protect the pile head. A 2,800-lb. hammer falling 20 ft. did not injure the piles. One pile was given 300 blows with a 2,800-lb. hammer falling 12 ft., and when pulled was unbroken. It was found that 30 ft. piles and under could be picked up safely by one end; longer piles cracked at the center when so handled. These long piles were successfully handled by a long chain, one end being wrapped around the pile at the center and the other end similarly wrapped near the head; the hook of the hoisting fall was hooked into the loop of the chain and as the pile was hoisted the hook slipped along the chain toward the top gradually up ending the pile. The piles weighed 175 lbs. per lin. ft. It was attempted to mold the piles directly on the ground by leveling it off and covering it with tar paper, but the ground settled and the method proved impracticable.

Method of Molding and Jetting Piles for Building Foundations.—In a number of foundations Mr. Frank B. Gilbreth has used a polygonal pile, either octagonal or hexagonal, with the sides corrugated or fluted as indicated in Fig. 61. In longitudinal section these piles have a uniform taper from butt to point and have flat points. Each pile is cored in the center, the core being 4 ins. in diameter at the top and 2 ins. at the bottom end. On each of the octagon or hexagon sides the pile has a half-round flute usually from 2½ to 3 ins. in diameter. The principal object of these flutes or "corrugations" is to give passage for the escape to the surface of the water forced through the center core hole in driving the pile. They are also for the purpose of increasing the perimeter of the pile and thereby gaining greater surface for skin friction.

The piles are reinforced longitudinally and transversely. On this particular job the reinforcement was formed with Clinton Electrically Welded Fabric, the meshes being 3 ins.×12 ins.; the longer dimension being lengthwise with the pile and of No. 3 wire; the horizontal or transverse reinforcement being of No. 10 wire. The meshes being electrically welded together, the reinforcement was got out from a wide sheet taking the form of a cone. No part of the reinforcement was closer than 1 in. from the outside of the concrete. In general only sufficient sectional area of material is put in the reinforcement to take the tensile stresses caused by the bending action when handling the pile preparatory to driving; more reinforcement than this only being necessary when the piles are used for wharves, piers or other marine structures, where a considerable length of pile is not supported sidewise or when they are subjected to bending stresses.

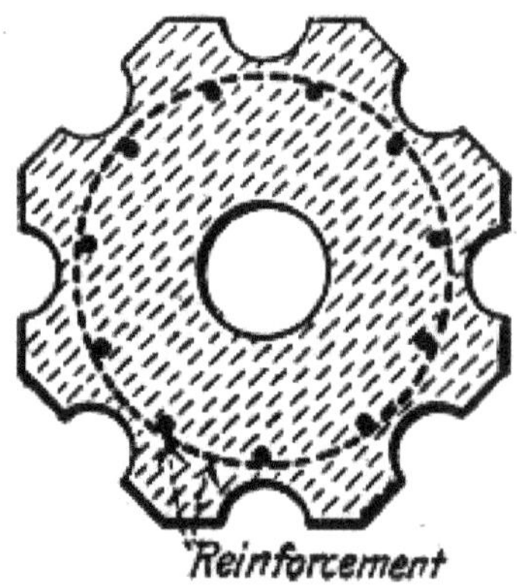

Fig. 61.—Cross-Section of Corrugated Reinforced Concrete Pile.

Molding.—The forms for molding the piles are made from 2-in. stuff, gotten out to the required dimensions, the corrugations being formed by nailing pieces on the inside whose section is the segment of a circle. The sides of the octagon are fastened to the ends through which the core projects some 6 or 8 ins. At times while the molding of the pile is in progress, the central core is given a partial turn to prevent the setting of the cement holding it fast and thereby preventing the final removal.

The stripping of the forms from the piles is usually done from 24 to 48 hours after molding, and from this time on great care is taken that there is a sufficient amount of moisture in the pile to permit of the proper action for setting of the cement. This is usually accomplished by covering the piles over with burlaps and saturating with water from a hose; the operation of driving the pile not being attempted until the concrete is at least ten days old.

Driving.—The operation of driving corrugated concrete piles is somewhat similar to that for driving ordinary wooden piles by water jet, but a much heavier hammer with less drop is used. The jetting is accomplished by inserting a 2-in. pipe within the pile. This pipe is tapered at the bottom end to 1-in. diameter, forming a nozzle, and the water pressure used is about 120 lbs. per sq. in. As a rule, this pressure is obtained by the use of a steam pump which may be connected with the boiler which operates the pile driver, or with a separate steam supply. At the upper end of this 2-in. pipe an elbow is placed and a short length of pipe is connected to this and to the hose from the water supply.

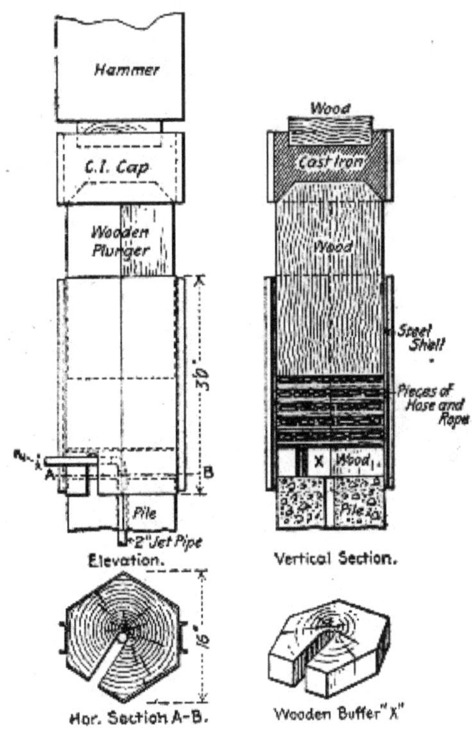

Fig. 62.—Cushion Cap for Driving Gilbreth Corrugated Pile.

As it is not practicable to drop the hammer directly on the head of the concrete piles, the driving is accomplished by the use of a special cap, Fig. 62. This cap is about 3 ft. in height and the bottom end fits over the head of the pile. In one side of this cap is a slot from the outside to the center, which permits the 2-in. pipe, which supplies the water jet for driving the pile, to project. The outside of this cap is formed with a steel shell, the inside has a compartment filled with rubber packing and the top has a wooden block which receives a blow from the hammer. In this way the head of the pile is cushioned, which prevents the blow of the hammer from bruising or breaking the concrete.

During the operation of driving, the water from the jet comes up on the outside of the pile and carries with it the material which it displaces in driving. This, with the assistance of the hammer, allows the pile to be driven in place, and, contrary to what might be supposed, after the operation of driving when the water has saturated into the ground or been drained away, this operation puddles the earth around the pile, so that after a few hours' time the skin friction is much more than it would be with the pile driven into more compact soil without the use of a jet.

Fig. 63.—View Showing Method of Fabricating Reinforcement for a Round Pile with Flattened Sides.

Method of Molding and Driving Round Piles.—In constructing a warehouse at Bristol, England, some 600 spirally-reinforced piles of the

Coignet type were used. Coignet piles are in section circles with two longitudinal flat faces to facilitate guiding during driving; this section is the same as would be found by removing two thin slabs from opposite sides of a timber pile. The reinforcement consists of longitudinal bars set around the periphery and drawn together to a point at one end and then inserted into a conical shoe; these longitudinal bars are wound spirally with a ¼-in. rod wire tied to the bars at every intersection. This spiral rod has a pitch of only a few inches, but to bind it in place and give rigidity to the skeleton it is wound by a second spiral with a reverse twist and a pitch of 4 or 5 ft. As thus constructed, the reinforcing frame is sufficiently rigid to bear handling as a unit. The piles used at Bristol were 14 to 15 ins. in diameter and 52 ft. long, and weighed about 4 tons gross each. The mixture used was cement, river sand and crushed granite.

Molding.—In molding Coignet piles the reinforcement is assembled complete as shown by Fig. 63 and then suspended as a unit in a horizontal mold constructed as shown by the cross-section Fig. 64. The concrete is deposited in the top opening and rammed and worked into place around the steel after which the opening is closed by the piece *A*. After 24 hours the curved side pieces *B* and *C* are removed and the pile is left on the sill *D* until hard enough to be shifted; a pile is considered strong enough for driving when about six weeks old.

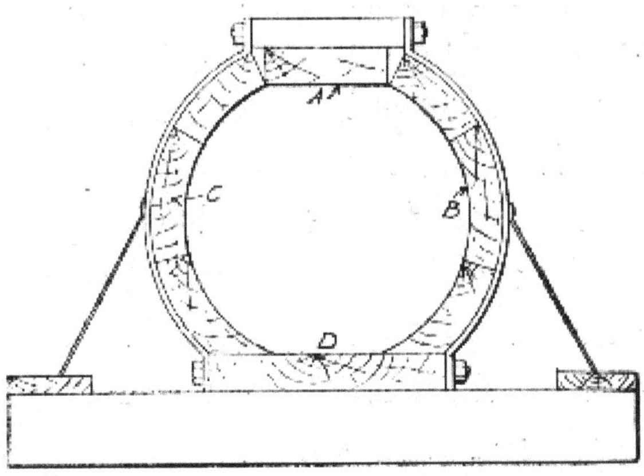

Fig. 64.—Form for Molding Round Pile with Flattened Sides.

Driving.—Coignet piles at the Bristol work were handled by a traveling crane. The material penetrated was river mud and they were driven with a hammer weighing 2 tons gross; in driving the pile head was encircled by a metal cylinder into which fitted a wooden plunger or false pile with a bed of shavings and sawdust between plunger and pile head.

Molding and Driving Square Piles for a Building Foundation.—The Dittman Factory Building at Cincinnati, O., is founded on reinforced concrete piles varying from 8 to 22 ft. in length. The piles were square in cross-section, with a 2-in. bevel on the edges; a 16-ft. pile was 10 ins. square at the point and 14 ins. square at the head, shorter or longer piles had the same size of point, but their heads were proportionally smaller or larger, since all piles were cast in the same mold by simply inserting transverse partitions to get the various lengths. Each pile was reinforced by four ¾-in. twisted bars, one in each corner, bound together by ¼-in hoops every 12 ins.. The bars were bent in at the point and inserted in a hollow pyramidal cast iron shoe weighing about 50 lbs. The concrete was a 1-2-4 stone mixture and the pile was allowed to harden four weeks before driving. They were cast horizontally in wooden molds which were removed after 30 hours.

Driving.—Both because of their greater weight and because of the care that had to be taken not to shatter the head, it took longer to adjust and drive one of these concrete piles than it would take with a wooden pile. The arrangement for driving the piles was as follows: A metal cap was set over the head of the pile, on this was set the guide cap having the usual wood deadener and on this was placed a wood deadener about 1 ft. long. The metal cap was filled with wet sand to form a cushion, but as the pile head shattered in driving the sand cushion was abandoned and pieces of rubber hose were substituted. With this rubber cushion the driving was accomplished without material damage to the pile head.

The hammer used weighed 4,000 lbs. and the drop was from 4 to 6 ft. The blows per pile ranged from 60 up. The average being about 90. In some cases where the driving was hard it took over 400 blows to drive a 14-ft. pile. An attempt to drive one pile with a 16-ft. drop resulted in the fracture of the pile.

Method of Molding and Driving Octagonal Piles.—The piles were driven in a sand fill 18 ft. deep to form a foundation for a track scales in a railway yard. They were octagonal and 16 ins. across the top, 16 ft. long, and tapered to a diameter of 12 ins. at the bottom. They were also pointed for about a foot. The reinforcement consisted of four ½-in. Johnson corrugated bars spaced equally around a circle concentric with the center of the pile, the bars being kept 1½ ins. from the surface of the concrete. A No. 11 wire wrapped around the outside of the bars secured the properties of a hooped-concrete column. The piles were cast in molds laid on the side. They were made of 1:4½ gravel concrete, and were seasoned at least three weeks before being driven.

An ordinary derrick pile driver, with a 2,500-lb. hammer falling 18 ft., was used in sinking them. A timber follower 6 ft. long and banded with iron straps at both ends was placed over the head of the pile to receive directly the hammer blows. The band on the lower end was 10 ins. wide and extended 6 ins. over the end of the follower. In this 6-in. space a thick sheet of heavy rubber was placed, coming between the head of the pile and the follower. Little difficulty was experienced in driving the piles in this manner, although 250 to 300 blows of the hammer were required to sink each pile. The driving being entirely through fine river sand there is every probability that any kind of piles would have been driven slowly. The heads of the first 4 or 5 piles were battered somewhat, but after the pile driver crew became familiar with the method of driving, no further battering resulted and the heads of most of the piles were practically uninjured.

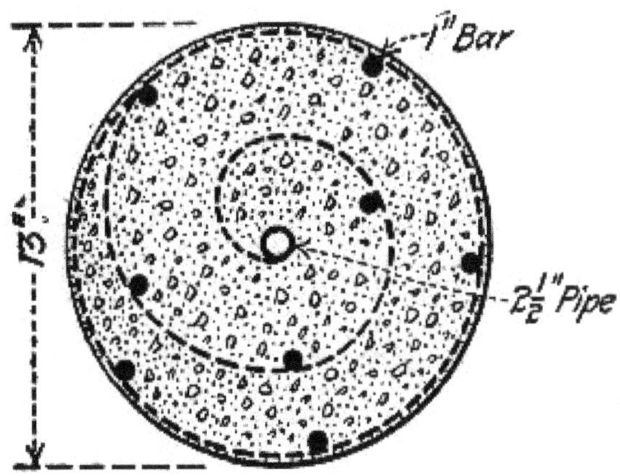

Fig. 65.—Cross-Section of Chenoweth Rolled Pile.

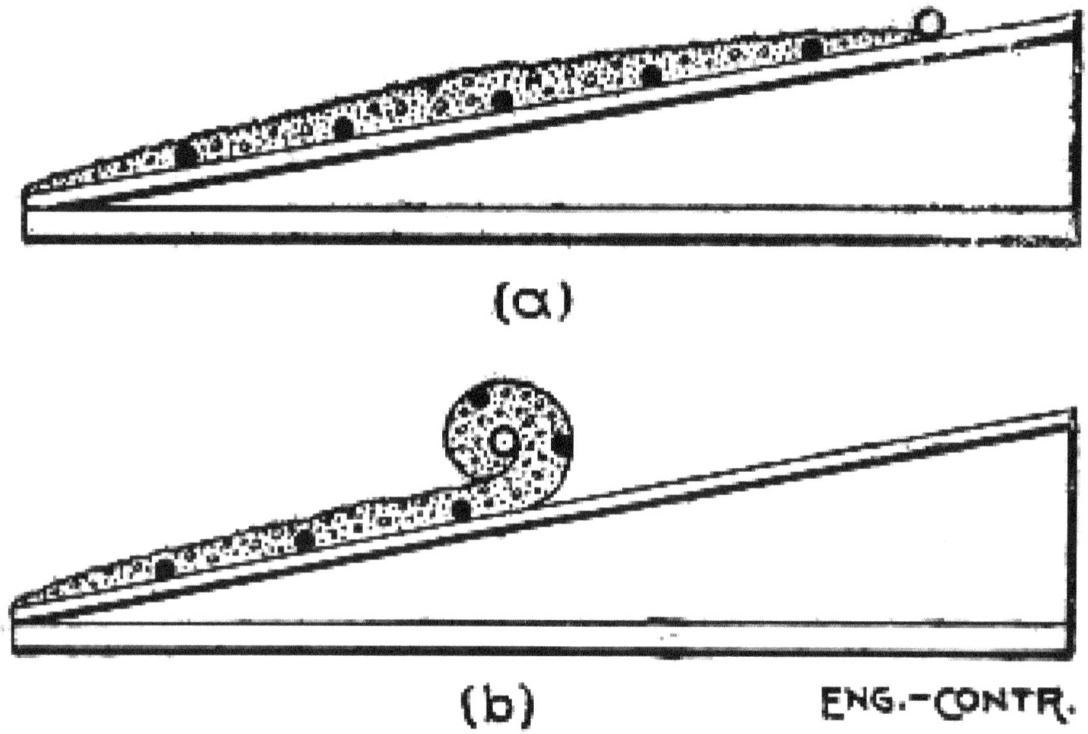

Fig. 66.—Diagram Showing Method of Rolling Chenoweth Pile.

Method and Cost of Making Reinforced Concrete Piles by Rolling.—In molding reinforced concrete piles exceeding 30 or 40 ft. in length, the problem of molds or forms becomes a serious one. A pile mold 50 or 60 ft. long is not only expensive in first cost, but is costly to maintain, because of the difficulty of keeping the long lagging boards from warping. To overcome these difficulties a method of molding piles without forms has been devised and worked out practically by Mr. A. C. Chenoweth, of Brooklyn, N. Y. This method consists in rolling a sheet of concrete and wire netting into a solid cylinder on a mandril, by means of a special machine. Fig. 65 is a sketch showing a cross-section of a finished pile, in which the dotted line shows the wire netting, the hollow circle is the gas pipe mandril, and the solid circles are the longitudinal reinforcing bars.

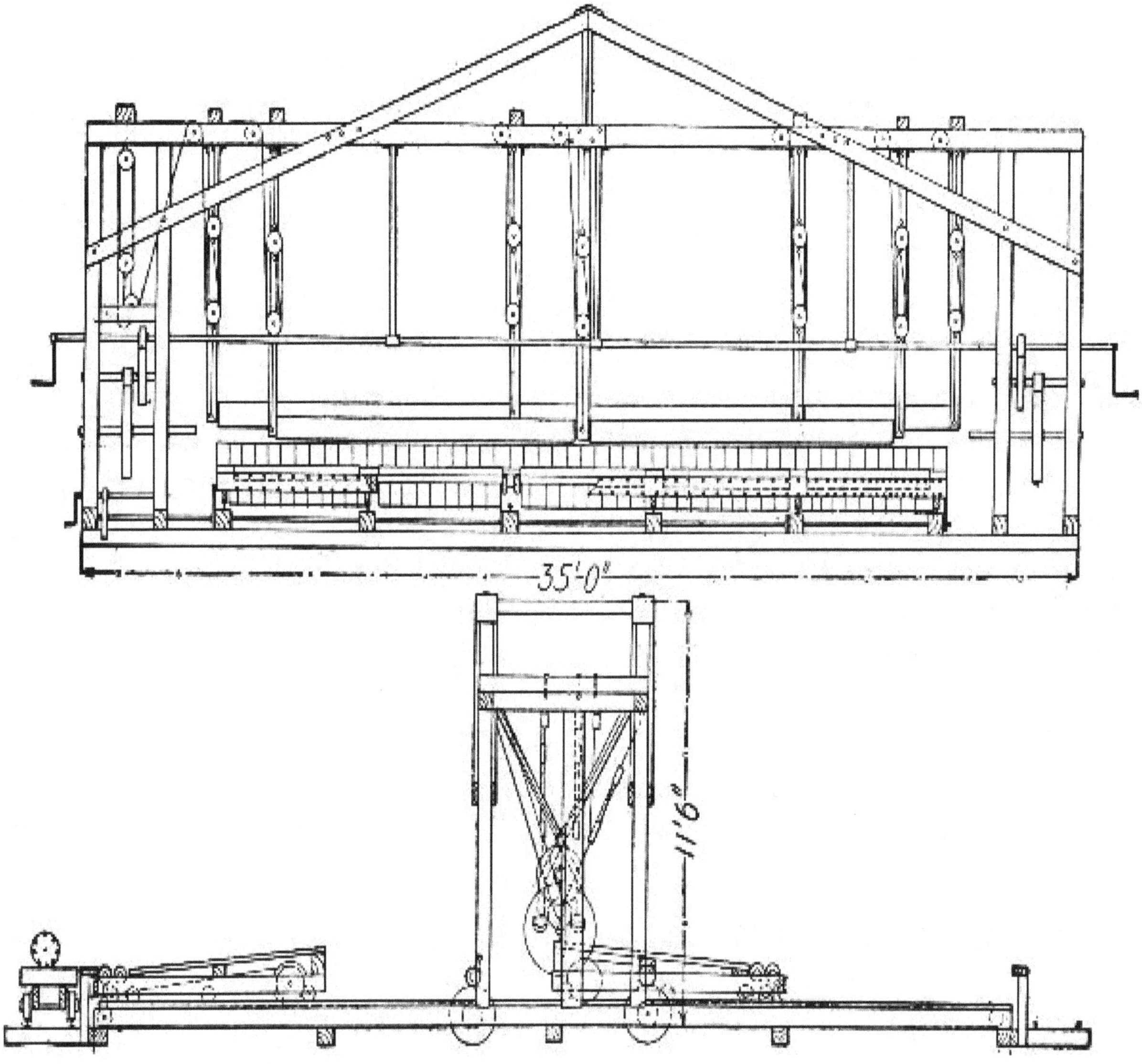

Fig. 67.—Machine for Rolling Chenoweth Piles.

In making the pile the netting is spread flat, with the reinforcing bars attached as shown at (*a*), Fig. 66, and is then covered with a layer of concrete. One edge of the netting is fastened to the platform, the other edge is attached to the winding mandril. The winding operation is indicated by sketch (*b*), Fig. 66. Fig. 67 shows the machine for rolling the pile. It consists of a platform and a roll. The platform is mounted on wheels and is so connected up that it moves back under the roll at exactly the circumferential speed of the roll; thus the forming pile is under constant, heavy pressure between the roll and platform. When the pile has been completely rolled it is bound at intervals by wire ties; the wire for these ties is carried on spools arranged under the edge of the platform at intervals of 4 ins. for the first 10 ft. from the point and of 6 ins. for the remainder of the length. The binding is done by giving the pile two or three extra revolutions and then cutting and tying the wire; then by means of a long removable shelf which contains the flushing mortar, as the pile revolves it becomes coated on the outside with a covering that protects the ties and other surface metal. Finally the pile is rolled onto a suitable table to harden.

An exhibition pile rolled by the process described is 61 ft. long and 13 ins. in diameter. This pile was erected as a pole by hoisting with a tackle attached near one end and dragging the opposite end along the ground exactly as a timber pole would be erected. It was also suspended free by a tackle attached at the center; in this position the ends deflected 6 ins. Neither of these tests resulted in observable cracks in the pile. The pile contains eight 1-in. diameter steel bars 61 ft. long, one 2½-in. pipe also 61 ft. long, 366 sq. ft., or 40.6 sq. yds. ½-in. mesh 14 B. & S. gage wire netting, and 2 cu. yds. loose concrete. Its cost for materials and labor was as follows:

Materials—	
Gravel, 28.8 cu. ft., at $1 per cu. yd.	$ 1.05
Sand, 19.8 cu. ft., at $1 per cu. yd.	.73
Cement, 3 bbls., at $1.60 per bbl.	4.80
Netting, 40.6 sq. yds., at 17½ cts. per sq. yd.	7.10
Rods, wire, etc., 1,826 lbs., at 2½ cts. per lb.	45.65
Total	$59.33
Mixing 2 cu. yds. concrete, four men one hour, at 15 cts. per hour	$ 0.60
Placing concrete and netting, four men 30 mins., at 15 cts. per hour	.30
Winding pile, four men 20 mins., at 15 cts. per hour	.20
Removing pile, four men 10 mins., at 15 cts. per hour	.10
	$1.20
Grand total	$60.53

This brings the cost of a pile of the dimensions given to about $1 per lin. ft.

11

HEAVY CONCRETE WORK

The construction problem in building concrete structures of massive form and volume is chiefly a problem of plant arrangement and organization of plant operations. In most such work form construction is simple and of such character that it offers no delay to placing the concrete as rapidly as it can be produced. The same is true of the character of the structure, it is seldom necessary for one part of the work to wait on the setting and hardening of another part. As a rule, there is no reinforcement to fabricate and place and where there is it is of such simple character as not to influence the main task of mixing, handling, and placing concrete. Stated broadly, the contractor in such work generally has a certain large amount of concrete to manufacture, transport and deposit in a certain space with nothing to limit the rapidity of these operations, except the limitations of plant capacity and management. Installation and operation of mixing and conveying plant, then are matters to be considered carefully in heavy concrete work.

In the following sections we have given one or more examples of nearly every kind of heavy concrete work excepting bridge foundations and retaining walls, which are considered in Chapters XII and XIII, and except rubble concrete work, which is considered in Chapter VI. In each case so far as the available records made it possible, we have given an account of the plant used and of its operation.

FORTIFICATION WORK.—Concrete for fortification work consists very largely of heavy platforms and walls for gun foundations and enclosures and of heavily roofed galleries and chambers for machinery and ammunition. The work is very massive and in the majority of cases of simple form. A large number of data are to be found in the reports of the Chief of Engineers, U. S. A., on all classes of fortification work, but the manner in which they are recorded makes close analysis of relative efficiencies of methods or of relative costs almost impossible. The following data are given, therefore, as examples that may be considered fairly representative of the costs obtained in fortification work done under the direction of army engineers; these data are not susceptible of close analysis because wages, working force, outputs, etc., are nearly always lacking.

Gun Emplacements, Staten Island, N. Y.—The work comprised 5,609 cu. yds. of concrete in two 12-in. gun emplacements, and 3,778 cu. yds. of concrete in two 6-in. gun emplacements. Concrete was mixed in a revolving cube mixer with the exception of 809 cu. yds. in the 6-in. emplacements which were mixed by hand at a cost of 56 cts. more per cubic yard than machine mixing cost. The body of the concrete was a 1-3-5 Portland cement, beach sand and broken trap rock mixture. The floors and upper surface of the concrete had a pavement consisting of 6 ins. of 1-3-5 concrete surfaced with 2 ins. of 1-3 mortar. Wages are not given, but for the time and place should have been about $1.50 per 8-hour day for common labor. The cost of materials was:

Alpha Portland cement, per bbl.	$1.98
Broken trap rock, per cu. yd.	0.81
12-in. emplacement, hauling sand per cu. yd.	0.175
6-in. emplacement, hauling sand per cu. yd.	0.20

The cost of the concrete in place was as follows:

Body Concrete——	12-in., per cu. yd.	6-in., per cu. yd.
Cement, at $1.98 per bbl.	$2.546	$2.546
Broken stone, at 81 cts. per cu. yd.	1.041	1.041
Sand, at 17½ and 20 cts. per cu. yd.	0.225	0.257
Receiving and storing materials at 11.6 cts. per cu. yd. and 8.4 cts. per bbl.	0.149	0.180
Mixing, placing and ramming	0.879	1.110
Forms, lumber and labor	0.477	0.950
Superintendence and miscellaneous	0.190	0.150
Total	$5.507	$6.234
Concrete Pavement—		
Materials	$2.97	$3.06
Labor	4.63	4.72
Total	$7.60	$7.78

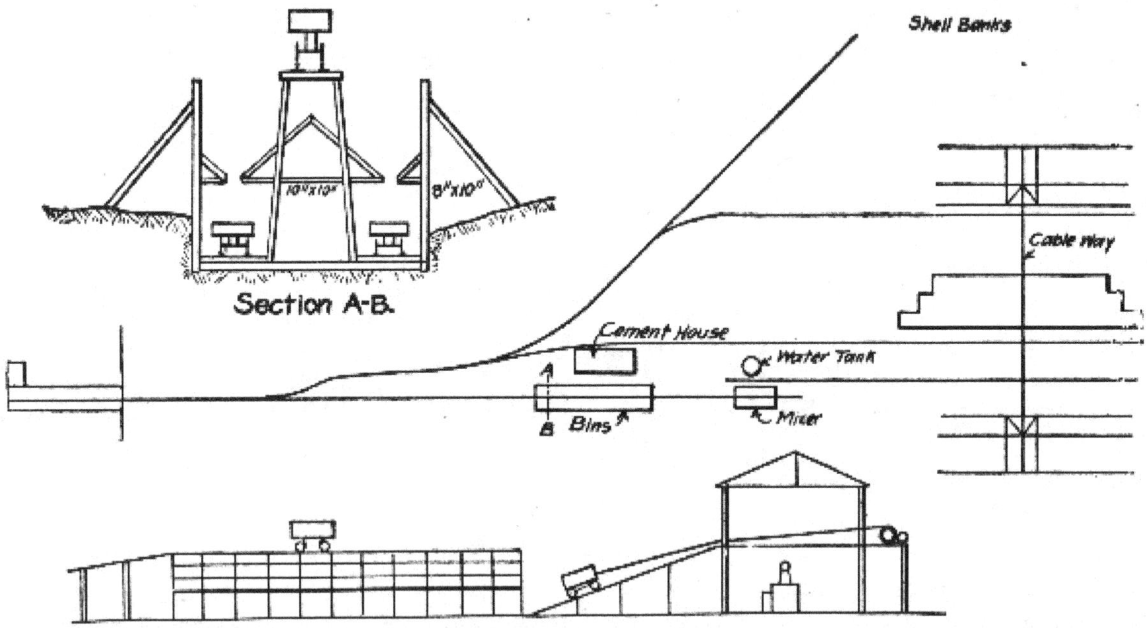

Fig. 68.—Sketch Plans of Concrete Making Plant for Mortar Battery Platform.

Mortar Battery Platform, Tampa Bay, Fla.—The platform contained 8,994 cu. yds. of concrete composed of a mixture of Portland cement, sand, shells and broken stone. The broken stone and cement were brought in by vessel and the sand and shells were obtained from the beach near by. The plant for the work was arranged as shown by the sketch, Fig. 68. Sand, stone and shells were stored in separate compartments in the storage bins. Box cars, divided into compartments of such size that when each was filled with its proper material, the car would contain the proper proportions for one batch of concrete, were pushed by hand under the several compartments of the bin in succession until charged; then they were hooked to a cable and hauled to the platform over the mixer and dumped. The charge was then turned over with shovels and shoveled into the hopper of a continuous mixer, located beneath. Two cars were used for charging the mixer, running on separate tracks as shown. The mixer discharged into buckets set on flat cars, which were hauled by mules under the cableway, which then lifted and dumped the bucket and returned it empty to the car. By using three bucket cars, one was always ready to receive the mixer discharge as soon as the preceding one had been filled, so that the mixer operated continuously. The cableway had a working span of 270 ft., the cable being carried by traveling towers 69 ft. high; the cableway was very easily operated back and forth along the work. The cableway complete, with 497 ft. of six-rail track for each tower, cost $4,700. The cost of materials and labor for the 8,994 cu. yds. of concrete was as follows:

	Per cu. yd.
1 bbl. cement at $2.46	$2.46
0.89 cu. yd. stone, at $2.95	2.622
0.315 cu. yd. shells, at $0.45	0.142
0.51 cu. yd. sand, at $0.12	0.062
Mixing and placing	0.693
Total	$5.979

The above batch tamped in place to 30 cu. ft., or 1-1/9 cu. yds., which gives the cost as follows:

	Per cu. yd.
Cost of concrete tamped in place	$5.381
Cost of form work	0.370
Total cost	$5.751

In the preceding prices of cement and stone, 59 cts. and 29 cts. per cubic yard, respectively, are included for storage. The costs of sand and shells are costs of screening and storing. Rough lumber for forms cost $10.25, and dressed lumber $12.75 per M. ft. B. M.

Emplacement for Battery, Tampa Bay, Fla.—The emplacement contained 6,654 cu. yds. of Portland cement, sand, shells and broken stone concrete. The plant arrangement is shown by Fig. 69. The sand and shells were got near the site, using an inclined cableway running from a 40-ft. mast near the mixer to a deadman at the shell bank. All the sand for the fill around the emplacement was obtained in the same way. The other materials were brought by vessel to a wharf, loaded by derrick onto cars operated by an endless cable, and taken to the work. The storage bins and mixing plant were operated much like those for the mortar battery work, previously described. A cube mixer was used, and the concrete was handled from it to the work by a crane derrick covering a circle of 100 ft. in diameter. The cost of materials and concrete was as follows:

Cement, plus 7 cts. for storage per bbl.	$ 2.532
Stone, plus 38 cts. for storage per cu. yd.	3.047
Shells, excavating and storage.	0.481
Sand, excavating and storage.	0.250
Lumber, rough per M. ft. B. M.	10.25
Lumber, dressed per M. ft. B. M.	12.75

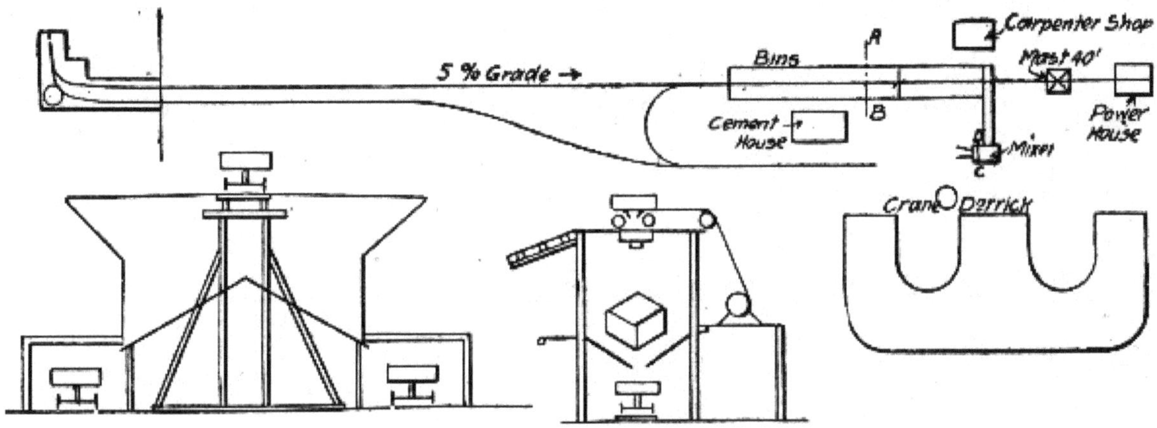

Fig. 69.—Sketch Plans of Concrete Making Plant for Battery Emplacement.

A batch made up as follows, tamped in place to a volume of 30 cu. ft. or 1-1/9 cu. yds.:

1 bbl. cement, at $2.532.	$ 2.532
0.315 cu. yd. shells, at $0.481.	0.151
0.51 cu. yd. sand, at $0.25.	0.130
0.89 cu. yd. stone, at $3.047.	2.710
Mixing and placing.	0.761
Total for 30 cu. ft.	$ 6.284

This gives a cost per cubic yard of concrete in place as follows:

Concrete in place, per cu. yd.	$ 5.655

Forms, per cu. yd. of concrete.	0.220
Total cost of concrete per cu. yd.	$ 5.875

United States Fortification Work.—The following methods and cost of mixing and placing concrete by hand and by cubical mixers is given by Mr. L. R. Grabill for U. S. Government fortification work done in 1899.

Hand Mixing and Placing.—The work was done by contract, using a 1 cement, 2 sand, 2 pebbles and 3 stone mixture turned four times. A board large enough for three batches at a time was used; one batch was being placed, one being mixed and one being removed at the same time so that the mixers moved without interval from one to the other. Two gangs were worked, each mixing 64 batches of 0.75 cu. yd., or 48 cu. yds. of concrete per day at the following cost:

Hand Mixing 9,000 Cu. Yds.—	Per day.	Per cu. yd.
6 men wheeling materials	$ 7.50	$0.16
8 men mixing	10.00	0.21
8 men wheeling away	10.00	0.21
6 men placing and ramming	7.50	0.16
1 pump man	1.25	0.02
1 waterboy	1.00	0.02
1 foreman	2.00	0.04
Totals	$39.25	$0.82

The entire cost of plant for this work was about $500.

Machine Mixing and Placing.—The concrete was mixed in a 4-ft. cubical mixer operated by a 12 hp. engine which also hauled the material cars up the incline to the mixer. These cars passed by double track under the material bins where the compartments of the car body were filled through trap doors; they then passed the cement house where the cement was placed on the load, then up the incline to the mixer and dumped, and then empty down an opposite incline. Seven turns of the mixer mixed the charge which was discharged into iron tubs on cars hauled by horses to two derricks whose booms covered the work. One gang by day labor mixed and placed 168 batches of 0.7 cu. yd., or 117.6 cu. yds. per day at the following cost:

Machine Mixing 4,000 Cu. Yds.—	Per day.	Per cu. yd.
32 men at $1.25	$40.00	$0.34
1 pumpman	1.25	0.01
1 teamster and horse	2.00	0.02
2 waterboys at $1	2.00	0.02
1 engineman	1.70	0.02
1 derrickman	1.50	0.01
1 fireman	1.50	0.01
1 foreman	2.88	0.03
Fuel (cement barrels largely)	1.25	0.01
Totals	$54.08	$0.47

The cost of the plant was about $5,000.

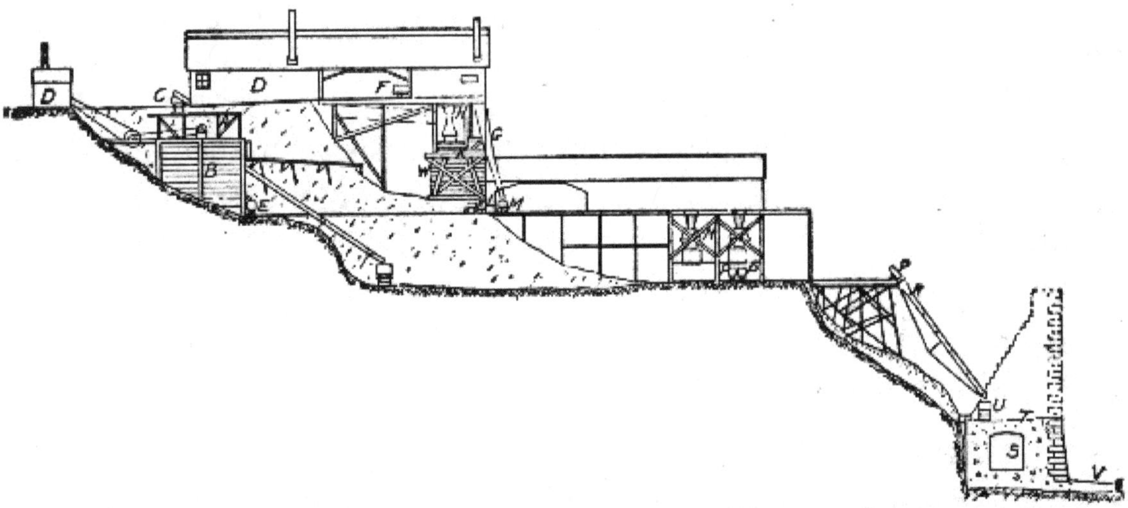

Fig. 70.—Concrete Making Plant for Constructing Lock Walls, Cascades Canal.

LOCK WALLS, CASCADES CANAL.—Four-fifths or 70,000 cu. yds. of lock masonry was concrete, the bulk of which was mixed and deposited by the plant shown by Fig. 70. The concrete was Portland cement, sand, gravel and broken stone. Cement was brought in in barrels by railway, stored and tested; from the store house the barrels were loaded onto cars and taken 250 ft. to a platform onto which the barrels were emptied and from which the cement was shoveled into the cement hopper and chuted to cars which took it to the charging hopper of the mixer. The stone was crushed from spalls and waste ends from the stone cutting yards, where stone for wall lining and coping and other special parts was prepared. These spalls and ends were brought in cars and dumped into the hopper of a No. 5 Gates crusher, with a capacity of 30 tons per hour. From the crusher the stone passed to a 2½-in. screen, the pieces passing going to a bin below and the rejections going to a smaller Blake crusher and thence to the bin. The dust and small particles were not screened out. The sand and gravel were obtained by screening and washing pit gravel. The gravel was excavated and brought in cars to the washer. This consisted of a steel cylinder 2 ft. 6½ ins. in diameter and about 18 ft. long, having an inclination of 1 in. per foot. An axial gudgeon supported the cylinder at the lower end and it rested on rollers at the other end and at an intermediate point. The gravel was fed by hopper and chute into the upper end and into this same end a 3-in. perforated pipe projected and extended to about mid-length of the cylinder. The cylinder shell was solid and provided with internal fins for about half its length from the feed end. For the remainder of its length nearly to the end, the shell was perforated with 2½-in. holes. For a length of 4 ft. beyond mid-point it was encircled by a concentric screen of ⅛-in. holes, and this screen for 3 ft. of its length was encircled by another screen of 30 meshes to the inch. The pit mixture fed into the cylinder was gradually passed along by the combined inclination and rotation, being washed and screened in the process. The sand fell into one bin and the gravel into another, and the waste water was carried away by a flume. The large stones passed out through openings at the lower end of the shell and were chuted into cars. The cars came to the mixer as clearly shown by Fig. 70.

The stone and gravel cars were side dump and the cement car was bottom dump. The mixers were of the cube type 4 ft. on each edge and operated by a 7×12-in. double cylinder engine at nine revolutions per minute. The usual charge was 32 cu. ft. of the several ingredients, and it was found that 15 revolutions requiring about 1½ minutes were sufficient for mixing. The average work of one mixer was 17 batches or about 13 cu. yds. per hour, but this could be speeded up to 20 batches per hour when the materials were freely supplied and the output freely removed. Two cars took the concrete from the mixer to the hopper, from which it was fed to the work by chute. The hopper was mounted on a truck and the chute was a wrought iron cylinder trussed on four sides and having a 45° elbow at the lower end to prevent scattering. The chute fed into a car running along the wall and distributing the material. It was found impracticable to move the chute readily enough to permit of feeding the concrete directly into place. As the concreting progressed upward the trestle was extended and the chute shortened. It was found that wear would soon disable a steel chute so that the main trussed cylinder had a smaller, cheaply made cylinder placed inside as a lining to take the wear and be replaced when necessary.

The plant described worked very successfully. Records based on 9,614.4 cu. yds. of concrete laid, gave the following:

	Cu. yds.
Concrete mixed by hand	1,777.0
Concrete mixed by machine	7,837.4
Total concrete laid	9,614.4
Concrete placed by derricks	2,372.0
Concrete placed by chute	7,242.4
Concrete 1-2-4 mixture	156.0
Concrete 1-3-6 mixture	1,564.0
Concrete 1-4-8 mixture	6,892.0

The average mixture was 1 cement, 3.7 sand, 4.8 gravel and 2.6 broken stone. The average product was 1.241 cu. yds. concrete per barrel of cement and 1.116 cu. yds. of concrete per cubic yard of stone and gravel. The average materials for 1 cu. yd. of concrete were: Cement 0.805 bbl., sand 0.456 cu. yd., gravel 0.579 cu. yd., and stone 0.317 cu. yd.

The cost of these 9,614.4 cu. yds. of concrete in place was:

Hand Mixed and Placed by Derrick—	Per cu. yd.
Labor mixing 1,777 cu. yds	$1,072
Repairs, fuel, etc	0.016

Total cost mixing	$1,088
Labor placing 2,372 cu. yds.	0.6025
Fuel, tramways, etc.	0.1958
Total cost placing	$0.7983

Machine Mixed and Placed by Chute—

Labor mixing 7,837 cu. yds.	$0.388
Repairs, fuel, etc	0.046
Total cost mixing	$0.434
Labor placing 7,242 cu. yds	0.414
Fuel, tramways, etc.	0.045
Total cost placing	$0.459

Materials and Supplies 9,614 cu. yds.—

Timbering	$0.145
Cement	3.289
Sand and gravel	1.073
Broken stone	0.536
Cement testing, repairs, etc.	0.223
Total	$5,266

Plant and Superintendence, 9,614 Cu. Yds.—

Engineering, superintendence, repairs, etc.	$1,508
20% cost of plant	0.165
Total	$1,673

The comparative cost of hand and machine mixing and handling was thus:

Item—	Hand.	Machine.

Mixing per cu. yd.	$1.088	$0.434
Placing per cu. yd.	0.798	0.459
Materials, etc., per cu. yd.	5.466	5.466
Plant, etc., per cu. yd.	1.673	1.right
Totals	$9.025	$8.032

The average total costs of all the concrete placed were:

Mixing per cu. yd.	$0.555
Placing per cu. yd.	0.543
Materials per cu. yd.	5.266
Plant, etc., per cu. yd.	1.673
Total	$8.037

LOCKS, COOSA RIVER, ALABAMA.—The following methods and costs are given by Mr. Charles Firth for constructing lock No. 31 for the Coosa River canalization, Alabama. This lock is 420 ft. long over all, 322 ft. between quoins, 52 ft. clear width, 14.7 ft. lift and 8 ft. depth of water on sills; it contained 20,000 cu. yds. of concrete requiring 21,500 bbls. cement, half Alsen and half Atlas.

Figure 71 shows the concrete mixing plant, consisting of two 4×4 ft. cube mixer, driven by a 10×16-in. engine. The top floor of the mixer house stored the cement, 2,000 bbls. The concrete was a 1-3-5½ stone mixture. Each mixer charge consisted of 3 cu. ft. cement, 9 cu. ft. sand and 16.5 cu. ft. stone; the charge was turned over four times before and six times after watering at a speed not exceeding eight revolutions per minute. The average output of the plant was 200 cu. yds. per 8-hour day, or 100 cu. yds. per mixer, but it was limited by the means for placing.

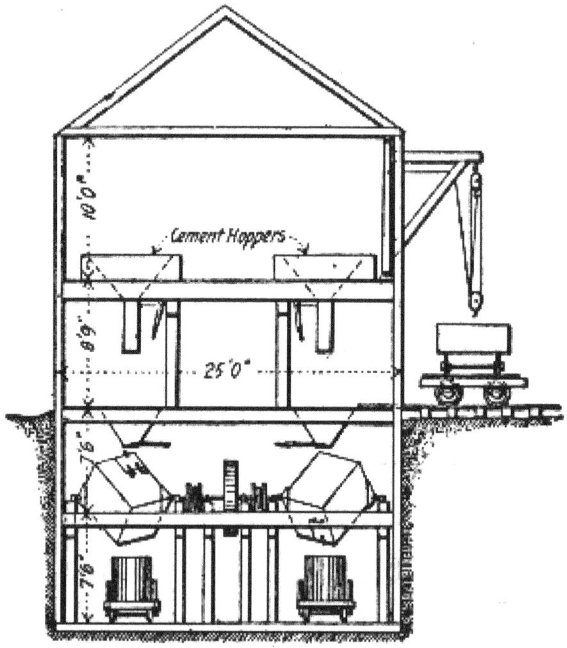

Fig. 71.—Concrete Mixing Plant for Lock Construction, Coosa River, Alabama.

The concrete was mixed dry, deposited in 6 to 8-in. layers, and rammed with 30-lb. iron rammers with 6-in. square faces. For all exposed surfaces a 6-in. facing of 1-3 mortar was placed by setting 2×12-in. planks 4 ins. from the laggings, being kept to distance by 2×4-in. spacers, placing and ramming the concrete behind them, then withdrawing them, filling the 6-in. space with mortar and tamping it to bond with the concrete. The walls were carried up in lifts, each lift being completed entirely around the lock before beginning the next; the first lift was 10.7 ft. high and the others 6 ft., except the last, which was 4.5 ft., exclusive of the 18-in. coping. The coping was constructed of separately molded blocks 3 ft. long, made of 1-2-3 concrete faced with 1-1 mortar and having edges rounded to 3 ins. radius.

In constructing the forms a row of 6×8-in. posts 24 ft. long and 5 to 7 ft. apart was set up along the inside of each wall and a similar row of posts 12 ft. long was set up along the outside. From the tops of the short posts 6×8-in. caps reached across the wall and were bolted to the long posts; these caps carried the stringers for the concrete car tracks. The lagging consisted of 3×10-in. planks dressed on all sides. The backs of the walls were stepped and as each step was completed the rear 12-ft. posts were lifted to a footing on its top and carried in the necessary distance. The front posts remained undisturbed until the wall was completed. The lagging was moved up as the filling progressed. As no tie bolts were permitted, these forms required elaborate bracing.

From the mixing plant, which was located on the bank above reach of floods, the concrete cars were dropped by elevator to the level of the track over the walls and then run along the wall and dumped onto platforms inside the forms and just below the track. This arrangement was adopted, because it was found that even a small drop separated the stone from the mortar. The concrete was shoveled from the platforms to place and rammed. The cars were bottom dumping with a single door hinged at the side; this door when swinging back struck the track stringers and jarred the form so that constant attention was necessary to keep it in line. It would have been much better to have had double doors swinging endwise of the car. Another point noted was that unless the track was high enough to give good head room at the close of a lift the placing and ramming were not well done.

The cost of 8,710 cu. yds. of concrete placed during 1895 by day labor employing negroes at $1 per 8-hour day was as follows per cubic yard:

1 bbl. cement	$2.48
0.88 cu. yd. stone at $0.76	0.67
0.36 cu. yd. sand at $0.34	0.12
Mixing, placing and ramming	0.88
Staging and forms	0.42
Total	$4.57

LOCK WALLS, ILLINOIS & MISSISSIPPI CANAL.—The locks and practically all other masonry for the Illinois & Mississippi Canal are of concrete. The following account of the methods and cost of doing this concrete work is taken from information published by Mr. J. W. Woermann in 1894 and special information furnished by letter. The decision to use concrete was induced by the fact that no suitable stone for masonry was readily available (the local stone was a flinty limestone, usually without bed, or, at best, in thin irregular strata, and cracked in all directions with the cracks filled with fire clay) while good sand and gravel and good stone for crushing were plentifully at hand. The concrete work done in 1893-4 comprised dam abutments, piers for Taintor gates and locks.

Dam Abutments.—Four dam abutments were constructed, three of which were L-shaped, with sides next to the river 40 ft. long and sides extending into the banks 20 ft. long; the top thickness was 3 ft., the faces were vertical and the backs stepped with treads of 14 to 16 ins., and the width of base was 0.4 of the height.

Each of these abutments was built in four 30-cu. yd. sections, each section being a day's work. The forms consisted of 2×8-in. planks, dressed on both sides, 2×8-in. studs spaced 2 ft. on centers and 4×6-in. braces. For the first two of the four abutments, the forms were erected in sections, the alternate sections being first erected and filled. When these sections had hardened the forms were shifted to the vacant sections and lined up to and braced against the completed sections. This method did not give well aligned walls, so in subsequent work the forms were erected all at once.

The concrete was mixed by hand. The sand and cement were mixed dry being turned four times and spread in a layer Pebbles and broken stone, previously wetted were spread over the sand and cement and the whole turned four times, the last turn being into wheelbarrows; about five common buckets of water were added during the mixing. The mixture sought was one that would ram without quaking. Two forms of rammers were used; for work next to forms a 4×6-in. rammer and for inside work 6-in diameter circular rammer weighing 20 lbs. The gang mixing and placing concrete consisted usually of:

Item.	Per Day.	Per Cu. Yd.
2 handling cement and sand	$ 3.00	$0.10
3 filling barrows with aggregate	4.50	0.15
8 mixing concrete	12.00	0.40
2 shoveling concrete into barrows	3.00	0.10
5 wheeling concrete to forms	7.50	0.25

1 spreading concrete	1.50	0.05
5 tamping concrete	7.50	0.25
Total, 26 men	$39.00	$1.30

These cubic yard costs are based on 30 cu. yds. of wall completed per 8-hour day. The cost in detail of two abutments containing 254 cu. yds. was per cubic yard as follows:

Item.	Per Cu. Yd.
1.65 bbls. Portland (Germania) cement	$ 5.60
0.5 cu. yd. crushed stone	2.07
0.24 cu. yd. gravel	0.59
0.53 cu. yd. sand	0.24
Lumber, forms, warehouses, platforms[D]	0.55
Carpenter work[E] ($9 per M.)	1.10
Mixing and placing	1.47
20 per cent. first cost of plant	0.31
Engineering and miscellanies	0.31
Total	$12.24

[D] Charging ¼ of first cost of $18 per M. ft.

[E] Carpenters $3.50, laborers $1.50 per day; there was one laborer to two carpenters.

The large amount of cement 1.65 bbls. per cubic yard was due to facing the abutments with 8 ins. of 1-2 mortar. The concrete in the body of the wall was 1 cement, 2 sand, 2 gravel and 2 broken stone mixture. A dry mixture was used and this fact is reflected in the cost of ramming, 25 cts. per cu. yd. The cost of mixing was also high.

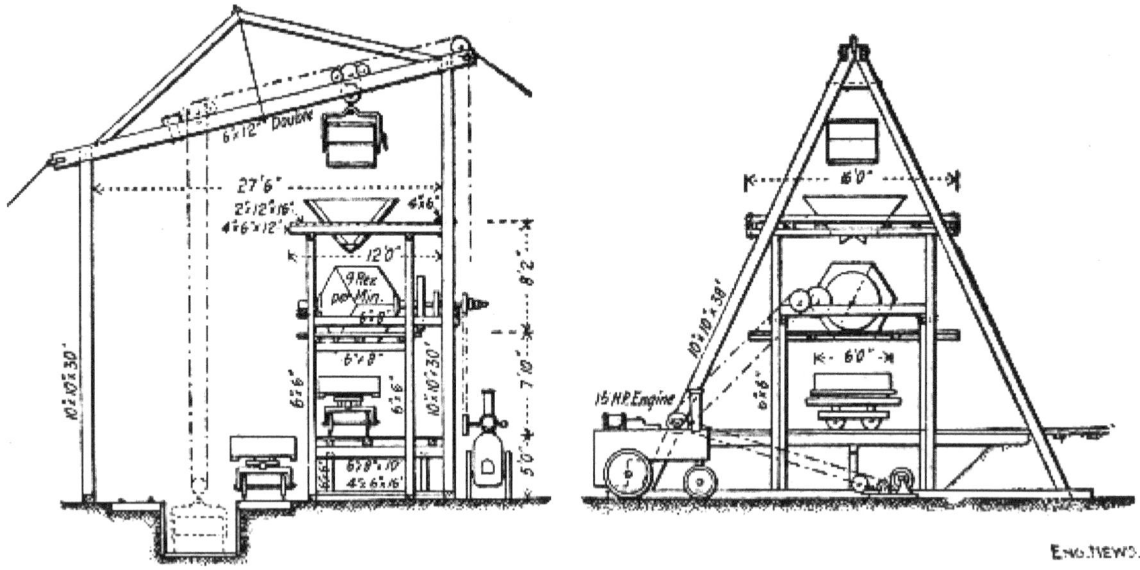

Fig. 72.—Concrete Mixing Plant for Lock Walls, Illinois & Mississippi Canal.

Piers for Taintor Gates.—The masonry at this point consisted of three piers 6×30 ft., and two abutments 30 ft. long, 6 ft. thick at base and 4 ft. thick at top, with wing walls; it amounted to 460 cu. yds. The feet of the inclined braces were set into gains in the horizontal braces and held by an 8-in. lag screw; after the posts were plumbed a block was lag-screwed at the upper end of each brace. These forms proved entirely satisfactory. The cost of the work per cubic yard was as follows:

Item.	Per Cu. Yd.
1.45 bbls. Portland cement	$4.330
0.55 cu. yd. crushed stone	0.604
0.252 cu. yd. pebbles	0.328
0.465 cu. yd. sand	0.419
40,000 ft. B. M. lumber (¼ cost of $16 per M.)	0.348
Carpenter work on forms	0.780
Mixing and placing concrete	1.909
20 per cent. cost of plant	0.090
Miscellaneous	0.182
Total	$8.99

Mixing Plant.—The concrete for all the lock work of 1893-4 was mixed by the plant shown by Figs. 72 and 73. The mixer plant proper consisted of a king truss carried by two A-frames of unequal height; under the higher end of the truss was a frame carrying a 4-ft. cubical mixer and under the lower end a pit for a charging box holding 40 cu. ft.

This charging box was hoisted by ½-in. steel cable running through a pair of double blocks as shown; the slope of the lower chord of the truss was such that the cable hoisted the box and carried it forward without the use of any latching devices. On two sides of the pit were tracks from the sand and stone piles and on the other two sides were the cement platform and water tank. The charging box dumped into the hopper above the mixer and the mixer discharged into cars underneath.

A 15-HP. engine operated the hoist by one pulley and the mixer by the other pulley. Nine revolutions of the mixer made a perfect mixture. The plant as illustrated was slightly changed as the result of experience in constructing the guard lock. The charging hopper was lowered 6 ins. and the space between the mixer and lower platform reduced by 9 ins.; diagonal braces were also inserted under the timbers carrying the mixer axles. This plant cost for framing and erection $300 and for machinery delivered $706. The crushing plant shown by Fig. 73 consisted of a No. 2 Gates crusher delivering to a bucket elevator.

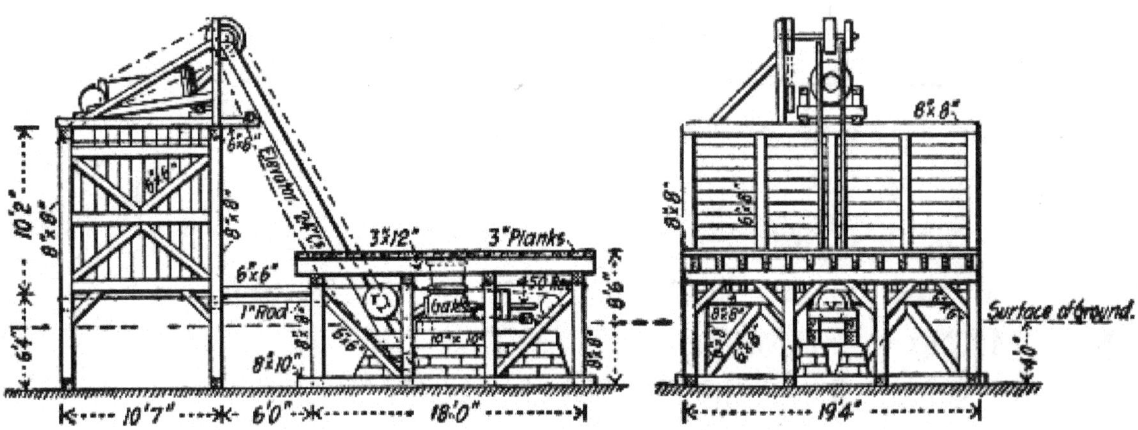

Fig. 73.—Stone Crushing Plant for Lock Walls, Illinois & Mississippi Canal.

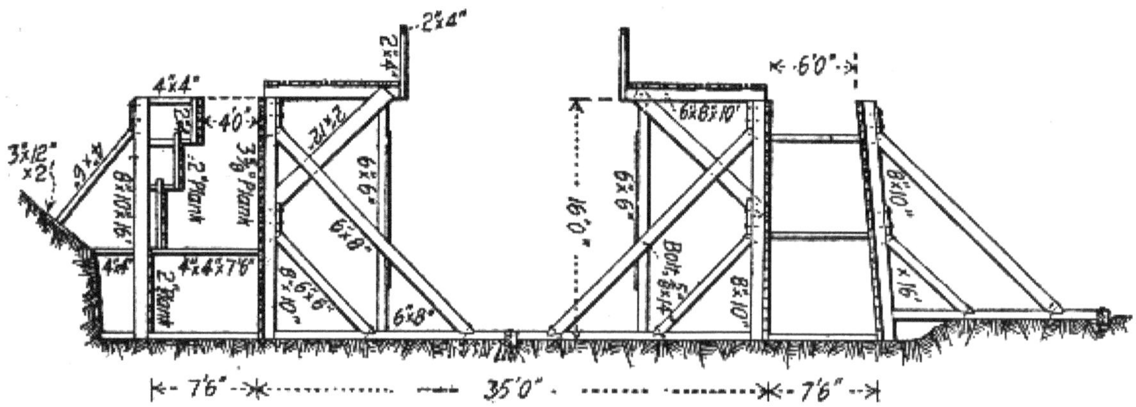

Section G-H.

Fig. 74.—Forms for Guard Lock, Illinois & Mississippi Canal.

Guard Lock.—The forms employed in constructing the guard lock are shown by Fig. 74, and in this drawing the trestle and platform for the concrete cars are to be noted. The walls were concreted in sections. A batch of concrete consisted of 1 bbl. cement, 10 cu. ft. sand and 20 cu. ft. crushed stone. The average run per 8-hour day was 40 batches of facing and 60 batches concrete, representing 100 bbls. cement. The gang worked was as follows:

Duty.	No. Men.	P. C. Cost.
Handling cement	3	5.26
Filling and pushing sand car	5	8.77
Filling and pushing stone car	9	15.79
Measuring water	1	1.75
Dumping bucket on top platform	3	5.26
Opening and closing door of mixer	1	1.75
Operating friction clutch	1	1.76
Attending concrete cars under mixer	1	1.76
Dumping cars at forms	2	3.51
Spreading concrete in forms	3	5.26
Tamping concrete in forms	10	17.54
Mixing mortar for facing	6	10.53
Finishing top of wall	2	3.51
Hauling concrete cars with 1 horse	1	3.51
Engineman operating hoist	1	3.51

Engineman operating engine	1	3.51
Foreman in charge of forms	1	3.51
General foreman	1	3.51
	—	——
Total	52	100.00

The percentages of cost in this statement have been calculated by the authors upon the assumption that each laborer received one-half as much wages as each engineman, foreman and horse and driver per 8 hours, which would make the total daily wages equivalent to the wages of 57 men. Wages of common labor were $1.50 per day.

Considering the size of the gang the output of 40 batches of mortar and 60 batches of concrete per day was very small. The total yardage of concrete in the guard lock was 3,762 cu. yds., 2,212 cu. yds. in the walls and 1,550 cu. yds. in foundations, culverts, etc. Its cost per cubic yard was made up as follows:

Item.	Total.	Per Cu. Yd.
5,246 bbls. Portland cement	$15,604}	$4.170
152 bbls. natural cement	84}	
2,910 cu. yds. stone	2,901	0.771
126 cu. yds. pebbles	113}	0.401
1,970 cu. yds. sand	1,398}	
145,000 ft. B. M. lumber (¼th cost)	659	0.175
Iron for forms, trestles, etc.	90	0.024
Coal, oil, miscellaneous	327	0.087
Carpenter work	2,726	0.724
Mixing and placing concrete	6,693	1.780
Pumping, engineering, misc.	742	0.197
20 per cent of plant	550	0.146
	——	——
Total	$31,887	$8.475

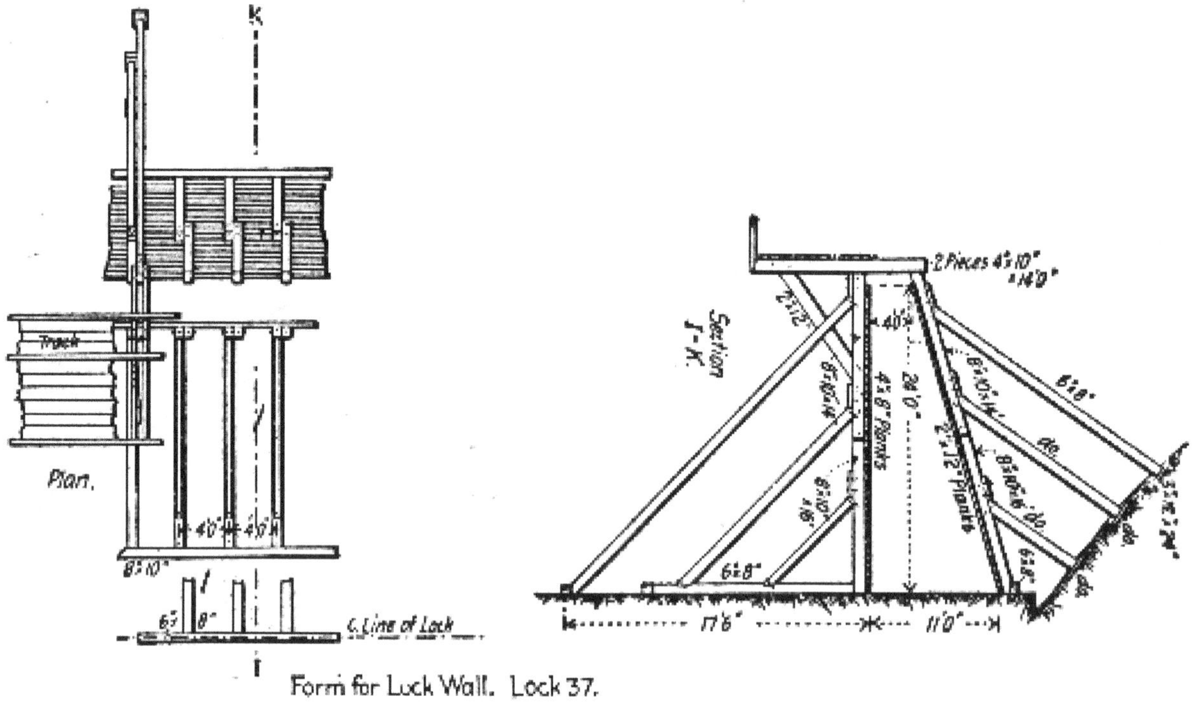

Fig. 75.—Forms for Regular Lock Walls, Illinois & Mississippi Canal.

Lock No. 37.—The character of the forms used in constructing the lock walls is shown by Fig. 75. The walls were built in sections and work was continuous with three 8-hour shifts composed about as specified for the guard lock work except that one or two men were added in several places making the total number 58 men. The average output per shift was 65 batches of concrete and 31 batches of facing mortar. The cost of the work, comprising 3,767 cu. yds., was as follows:

Item.	Total.	Per Cu. Yd.
4,564 bbls. Portland cement	$14,181	$3.764
2,460 cu. yds. crushed stone	4,521	1.200
250 cu. yds. pebbles	325	0.086
1,750 cu. yds. gravel	2,335	0.619
450 cu. yds. sand	450	0.119
180,000 ft. B. M. lumber (¼th cost)	990	0.236
Fuel, light, repairs, etc.	1,171	0.311
Carpenter work	2,526	0.671
Pumping	270	0.071

Mixing and placing concrete	6,170	1.632
20% cost of plant	730	0.193
Total	$33,669	$8.902

Lock No. 36.—The forms used were of the construction shown by Fig. 75. Three shifts were worked, each composed as specified for the guard lock, except that the number of tampers and spreaders was doubled, bringing the gang up to 65 men. The average output per gang per shift was 76 batches of concrete and 35 batches of facing mortar. The cost of 2,141 cu. yds. concrete in this lock was as follows:

Item.	Total.	Per Cu. Yd.
3,010 bbls. Portland cement	$9,057	$4.23
1,377 cu. yds. broken stone	1,922	0.90
393 cu. yds. pebbles	354	0.17
459 cu. yds. gravel	310	0.15
500 cu. yds. sand	889	0.42
150,000 ft. B. M. lumber (¼th cost)	600	0.28
Fuel, light, repairs, etc.	253	0.68
Carpenter work	1,472	0.11
Mixing and placing concrete	3,897	1.82
20% cost of plant	650	0.30
Total	$19,404	$9.06

The preceding data, made public by Mr. Woermann in 1894, are supplemented by the following information prepared for the authors:

"If any criticism was to be made of the concrete masonry erected in 1893 and 1894, it would probably be to the effect that it was too expensive. The cost of the masonry erected during those two seasons was $8 to $9 per cu. yd. Our records showed that about 45 per cent. of this cost was for Portland cement alone, and moreover, that 40 per cent. of the total cement used at a lock was placed in the 8-in. facing and 5-in. coping. So in the seven locks erected in 1895 on the eastern section, the facing was reduced to 3 ins. and the proportions changed from 1-2 to 1-2½.

"In 1898 this cost received another severe cut, and Major Marshall's instructions stated that the facing should not exceed 1½ ins. in thickness nor be less than ¾-in., while the layer of fine material on top of the coping was to be only sufficient to cover the stone and gravel. The amount of sand was again increased so that the proportions were 1-3.

"The cost of the Portland cement concrete was likewise cheapened by increasing the amount of aggregates. On the earlier work the proportions were 1-2-2-3, while on the work in 1898 the proportions were 1-4-4.

The cost of the walls was further cheapened by using Utica cement in the lower steps of the wall, with 2 ft. of Portland cement concrete on the face. The proportions used in the Utica cement concrete were 1-2½-2½. This lower step is one-third of the height, or about 7 ft.

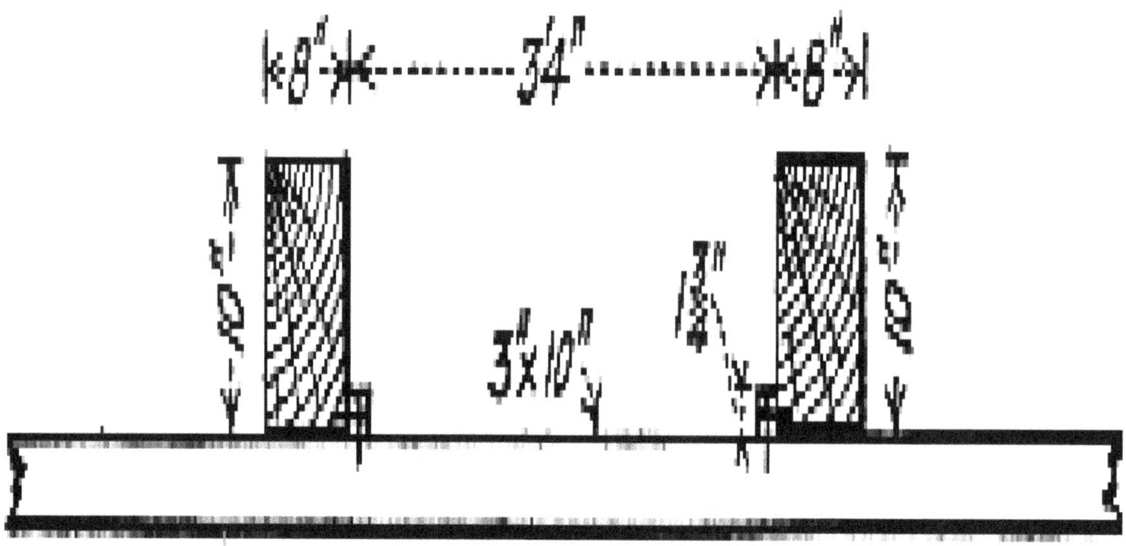

Fig. 76.—Sketch Showing Method of Attaching Lagging to Studs, Illinois & Mississippi Canal.

"The forms were of the same character as those used on the first locks, except that for lining the inner face, 3×10-in. hard pine planks were substituted for the 4×8-in. white pine. The hard pine was damaged less by the continuous handling, and the cost was practically the same.

There was also an important change made in the manner of fastening the plank to the 8×10-in. posts. A strip 1¾ ins. square was thoroughly nailed to each post, once for all, with 20d. spikes, and the planking was then nailed from the outside, as shown in Fig. 76.

This kept the face of the plank in a perfectly smooth condition, and prevented the formation of the little knobs on the face of the concrete which represented all the old nail holes. This style of forming was also easier to take apart after the setting of the concrete. Rough pine planks, 2×12-in., were used for the back of the form, the same as before.

"In order to keep ahead of the concrete force it was necessary to use two gangs of carpenters, erecting the forms for the next two locks. Each gang consisted of about 20 carpenters (at $2.25) and 10 helpers (at $1.50); but men were transferred from one to the other, according to the stage of completion of the two locks. In addition to these two gangs, two carpenters were on duty with each concrete shift to put in the steps in the back of the forms. Sufficient lumber was required for the forms for three complete locks, and 14 locks (Nos. 8 to 21) were built.

"The same type of mixer has been used as on the earlier work at Milan, namely, a 4-ft. cubical steel box mounted on corners diagonally opposite. On account of the greater number of locks to be built on the eastern section, however, two mixers were found necessary, so that while the concrete force was at work at one lock, the carpenters and helpers were erecting the mixer at the next lock. The facing was mixed by hand.

After turning over the dry cement and sand at least twice with shovels, the mixture was then cast through a No. 5 sieve, after which the water was incorporated slowly by the use of a sprinkling can so as to avoid washing. The secret of good concrete, after the selection of good materials, is thorough mixing and hard tamping. Each batch of concrete, consisting of about 1.2 cu. yds. in place, was turned in the mixer for not less than 2 mins. at the rate of 9 revolutions per minute. The amount of tamping is indicated by the fact that about 16 men out of 72 on each shift did nothing but tamp. The rammers used were 6 ins. square and weighed 33 lbs. The bottom of the rammer consisted of three ridges, each 1-in. in height, so as to make more bond between the successive layers.

"On the eastern section the top of the lock walls was higher above the ground, as a rule, than at the Milan locks, and the cars were run up an incline with a small hoisting engine. A 15-HP. portable engine and boiler operated the bucket hoist from one pulley, the mixer from the other pulley' and also furnished steam for the hoist which pulled the cars up the incline. The incline made an angle of about 30° with the ground. The practice of carrying on two sections at once was continued the same as on the western section. Each main wall was systematically divided into 11 sections, making each section about 20 ft. long. The corners of the coping were dressed to a quadrant of about 3 ins. radius with a round trowel like those used on cement walks. In fact, the whole method of finishing the coping was the same as is used on concrete walks. The mortar was put on rather wet and then allowed to stand for about 20 mins. before finishing. This allowed the water to come to the surface and prevented the formation of the fine water cracks which are sometimes seen on concrete work. After its final set the coping was covered with several inches of fine gravel which was kept wet for at least a week.

"The last concrete laid during the season was in November, on Lock No. 21, and Aqueducts Nos. 2 and 3. Portions of these structures were built when the temperature was below freezing. The water was warmed to about 60° or 70° F., by discharging exhaust steam into the tank. Salt was used only in the facing, simply sufficient to make the water taste saline. The maximum amount used on the coldest night when the temperature was about 20° F was 1½ per cent.

The concrete force on each shift was as follows:

Men.

Filling and pushing stone car	10
Filling and pushing gravel car	8
Measuring cement	3
Measuring water and cleaning bucket	2
Dumping bucket on top platform	2
Operating mixer	2
Loading concrete cars	1
Pushing and dumping cars on forms	3
Switchmen on forms	2
Spreading concrete in forms	12
Tamping concrete in forms	16
Mixing facing	3
Water boys	2
	—
Total laborers	66
Operating hoists	2
Finishing coping	2
Fireman	1
Sub-overseers	2
Overseer	1
	—
Total force	74

The cost of material and labor at Lock No. 15 (10-ft. lift), which contains 2,559 cu. yds. of concrete, was as follows:

Materials.	Per cu. yd.
0.56 bbl. Portland cement (0.96 per cu. yd.)	$1.42
0.64 bbl. Utica cement (1.58 per cu. yd.)	.30
0.58 cu. yd. stone	1.15
0.60 cu. yd. gravel	.52
14 ft. B. M. lumber[F] at $15 per M.	.21
0.6 lb. spikes	.01

Coal (10 tons in all, at $1.70)	.01
0.35 gal. kerosene	.03
Total materials	$3.65
Labor.	
Erecting forms ($7 per M.)	.45
Removing forms ($2 per M.)	.13
Erecting and removing mixer ($161)	.06
Loading and unloading materials at yards and lock sites	.23
Track laying ($86)	.03
Train service (narrow gage road)	.09
Delivering materials to mixer	.28
Mixing concrete	.11
Depositing concrete	.21
Tamping concrete	.21
Mixing, depositing and tamping, 69 cu. yds. face mortar ($160)	.23
General construction ($553)	.22
Total labor	$2.25

[F] The lumber was used nearly five times, which accounts for its low cost per cu. yd.

There were 1,430 cu. yds. of Portland cement concrete. 69 cu. yds. of Portland cement mortar facing, and 1,059 cu. yds. of Utica cement concrete. The Portland concrete cost $6.43 per cu. yd.; the Utica concrete, $4.77 per cu. yd. The following is the cost of labor on Lock No. 20 (11-ft. lift.; 2,750 cu. yds.):

	Per cu. yd.
Erecting forms ($7 per M.)	$.434
Removing forms ($1.70 per M.)	.113
Erecting and removing mixer ($151)	.058
Loading and unloading at yards, lock sites, etc.	.614
Tracks	.024

Train service (narrow gage)	.016
Pumping	.114
Delivering material to mixer	.288
Mixing concrete	.134
Depositing concrete	.205
Tamping concrete	.192
Mixing, depositing and tamping, 85 cu. yds. face mortar	.071
General construction	.246
Total	$2.509

COST OF HAND MIXING AND PLACING, CANAL LOCK FOUNDATION.—Mr. Geo. P. Hawley gives the following record of mixing and placing 4,000 cu. yds. of 1-4½ gravel concrete for the foundation of a lock constructed for the Illinois and Mississippi Canal in 1897. The concrete was mixed on 14×16-ft. board platforms, from which it was shoveled directly into place. The materials were brought to the board in wheelbarrows. Two boards were used, the usual gang for each being 4 men wheeling gravel, 4 men mixing, 1 man sprinkling, 2 men depositing and leveling and 2 men tamping. The two gangs were worked against each other. Ten hours constituted a day's work, and the average time and cost per cubic yard for mixing and placing were:

	Cts.
Foreman, 0.21 hr., at 30 cts	6.30
Laborers, 3.339 hrs., at 15 cts	50.09
Pump runner, 0.129 hr., at 20 cts	3.58
Water boy, 0.087 hr., at 7½ cts	0.65
Total labor per cu. yd., cents.	60.62

BREAKWATER AT MARQUETTE, MICH.—The breakwater extends out from the shore and consists of a prism of concrete resting on timber cribs filled with stone. Originally the cribs carried a timber superstructure; this was removed to give place to the concrete work. A typical cross-section of the concrete prism is shown by Fig. 77; the prism is 23 ft. wide on the base. Farther in shore the base width was reduced to 20 ft., and in the shore section the prism was changed to a triangular trapezoid by continuing the first slope to the bottom cutting off the berm and second slope. The wooden structure was removed to a level 1 ft. below mean low water and on it a concrete footing approximately 2 ft. thick was constructed for the prism proper. This footing reached the full width of the crib and was constructed in various ways during the 5 years through which the work continued. At first the footing concrete was deposited loose under water by means of bottom dumping buckets; later the stone filling of the cribs was simply leveled up by depositing concrete in bags, and last toe and heel blocks were molded and set flush with the sides of the crib and filled between. Methods of construction and records of cost are reported for portions only of the work and these are given here.

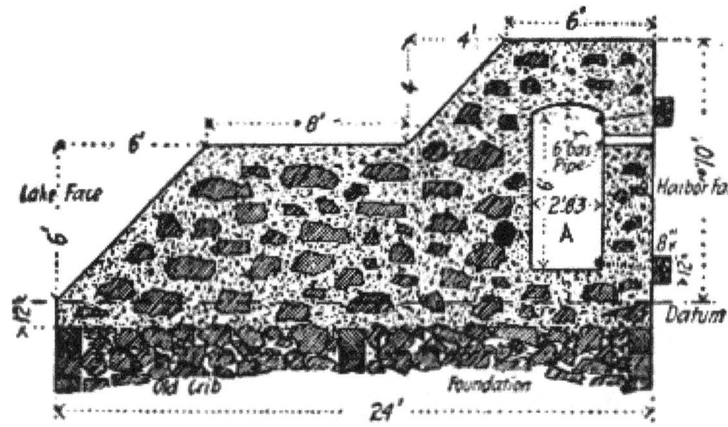

Fig. 77.—Cross Section of Marquette Breakwater.

Footing Placed under Water with Buckets.—Besides the material track which was constructed along the old wooden structure the plant consisted of a mixing scow and a derrick scow, which were moored alongside the work. The sand, stone and cement were brought out in cars between working hours and stored on the mixing scow, enough for one day's work at a time. The derrick handled a 40-cu. ft. bottom dump bucket, which sat in a well on the mixing scow with its top flush with the deck. The concrete was mixed by hand on the deck and shoveled into the bucket; the bucket was then handled by the derrick to the crib and lowered and dumped under water. The gang consisted of 24 men, 1 foreman, 1 master laborer, 14 men shoveling and mixing, 3 men wheeling materials, 1 derrick man and 3 men placing and depositing concrete. No record of output of this gang is available. The cost of the concrete in place with wages $1.25 to $1.40 per day for common labor is given as follows:

Materials.	Per cu. yd.
1.21 bbls. (459 lbs.) cement at $2.20	$2.657
1 cu. yd. stone at $1.58	1.580
0.5 cu. yd. sand at $0.50	0.250
2.02 lbs. burlap at $0.037	0.075
Twine and needles	0.005
Total materials	$4.567
Labor.	
Loading scow with materials	$0.4114
Mixing concrete	0.8459
Depositing concrete	0.5242
Total labor	$1.7815
Grand total	$6.348

These figures are based on some 757 cu. yds. of concrete footing. In explanation of the items of burlap, etc., it should be said that the cribs were carpeted with burlap to prevent waste of concrete into the stone fill.

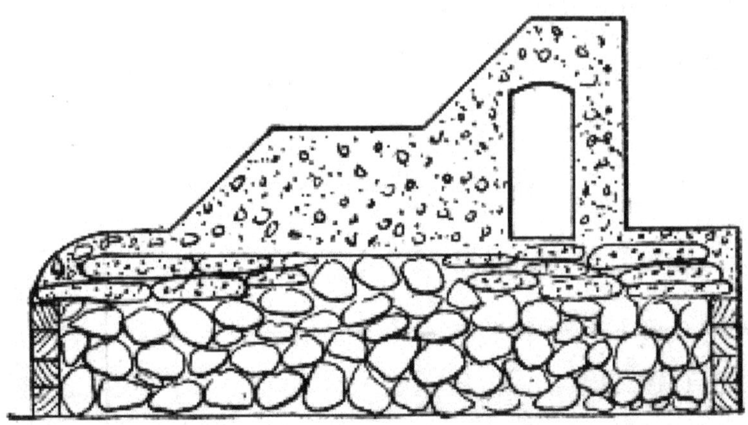

Fig. 78.—Cross Section of Marquette Breakwater Showing Manner of Constructing Footing with Bags of Concrete.

Leveling Off Cribs with Concrete in Bags.—The sketch, Fig. 78, shows the method of leveling off the cribs with concrete in bags. The concrete was mixed by hand on shore and filled into 8-oz. burlap bags, 6 ft. long and 80 ins. around, holding 2,000 lbs. The bags were filled while lying in position in a skip holding one bag. A skip was lifted by gallows frame and tackle onto a car and run out to the work where the derrick scow handled the skip to the crib, lowered it into the water and dumped the bag. The cost of making and placing some 375 cu. yds. of concrete in bags is given as follows:

Materials.	Total.	Per cu. yd.
453 bbls. cement at $2.627	$1,190.03	$3.173
375 cu. yds. stone at $1.619	607.13	1.619
180 cu. yds. sand at $0.392	70.56	0.188
3,220 yds. burlap at $0.03304	106.39	0.283
Twine and needles	6.36	0.017
Total materials	$1,980.47	$5.280
Labor Mixing.		
108 hrs. master laborer at $0.21⅞	$23.42	$0.062
1,750 hrs. labor at $0.175	306.25	0.816
Superintendence	12.55	0.033
Total labor mixing	$ 342.22	$0.911
Labor transporting.		
306 hrs. labor at $0.175	$53.55	$0.142
Superintendence	5.25	0.014
Total labor transporting.	$58.80	$0.156
Labor Depositing.		
108 hrs. engineman at $0.25	$27.00	$0.072
108 hrs. master laborer at $0.21⅞	23.42	0.062
510 hrs. labor at $0.175	89.25	0.238

Superintendence	13.25	0.035
Total labor depositing	$ 152.92	$0.407
Grand total labor	$ 553.94	$1.477
Grand total materials and labor	$2,534.41	$6.757

Molding Footing Blocks.—The blocks used at the toe of the prism were of the form and dimensions shown by Fig. 79. They were molded in a temporary shed heated to 50° to 65° F., and provided with a 2×8-in. dressed plank floor on 12×12-in. sills. `

The floor formed the bottoms of the block molds. Four molds were used, each consisting of four sides. Three laborers molded one block, 2.22 cu. yds. per day, wheeling, mixing, erecting and removing forms, placing concrete and doing all other work. The cost of making 40 blocks was recorded as follows:

Materials.	Total.	Per cu. yd.
126 bbls. cement at $2.75	$346.50	$3.893
88.9 cu. yds. screenings at $1.10	97.79	1.098
40.1 cu. yds. sand at $0.45	18.04	0.203
5 gals. oil at $0.65	3.25	0.036
Total materials	$465.58	$5.230
Labor.		
1,000 hrs. labor at $0.125	$125.00	$1.404
Watchman	29.15	0.327
Labor cutting wood for fuel	23.80	0.267
Superintendence	42.66	0.480
Total labor	$220.61	$2.478
Total labor and materials	$686.19	$7.708

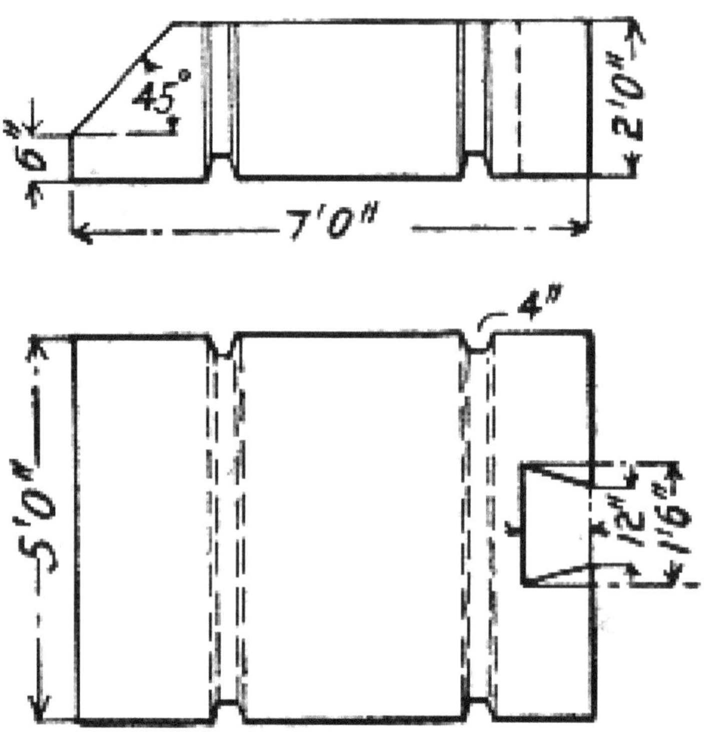

Fig. 79.—Details of Toe Blocks for Footing, Marquette Breakwater.

Molding Concrete Prism in Place.—The concrete prism was molded in alternate sections 10 ft. long; the form for the isolated sections consisted of eight pieces so constructed that when assembled in place and secured with bolts and turnbuckles the form was self-contained as to strength and required no outside support or bracing. The form once in place, all that remained to be done was to fill it, the block with the gallery through it being molded in one operation. The forms for the connecting blocks consisted of two slope panels, a panel for the harbor face and the gallery form, the blocks previously molded making the other sides of the form. The concrete was mixed by hand on shore, conveyed to the work in 1 cu. yd. cars and shoveled into the forms, where it was rammed with 35-lb. rammers. The following record covers 1,231 cu. yds. of concrete prism. In this concrete some 214 cu. yds. of rubble stone were embedded. The costs given are as follows:

Materials—	Total.	Per cu. yd.
1,780 bbls. natural cement at $1.068	$1,901.04	$1.545
963½ cu. yds. stone at $1.619	1,559.91	1.267

53½ cu. yds. screenings at $0.392	20.97	0.017
485.6 cu. yds. sand at $0.392	190.36	0.154
Miscellaneous materials	78.15	0.063
Totals	$3,750.43	$3.046
Labor Mixing—		
254 hrs. master laborer at $0.21⅞	$55.56	$0.045
4,470 hrs. labor at $0.175	782.42	0.635
Superintendence	18.20	0.015
Total labor mixing	$ 856.18	$0.695
Labor Transporting and Placing—		
35 days overseer at $2.33⅓	$81.67	$0.066
1,949 hrs. labor at $0.175	342.07	0.277
Superintendence	34.98	0.028
Total labor transporting and placing	$ 458.72	$0.371
Grand total, labor	$1,314.90	1.066
Total labor and materials	$5,065.33	4.112

No charge is made under materials for rubble stone as the only cost for this was cost of handling and this is included in transporting and placing.

BREAKWATER, BUFFALO, N. Y.—The following methods and costs of mixing and placing some 2,561 cu. yds. of concrete are given by Mr. Emile Low, for 10 parapet wall sections and 17 parapet deck sections for a breakwater at Buffalo, N. Y.

The concrete used was a 1 cement, 1 gravel, 1 sand grit and 4 unscreened broken stone. One bag of cement was assumed to measure 0.9 cu. ft. The voids in the sand grit and gravel were 27 per cent. and in the unscreened stone 39 per cent. The hardened concrete weighed 152 lbs. per cu. ft.

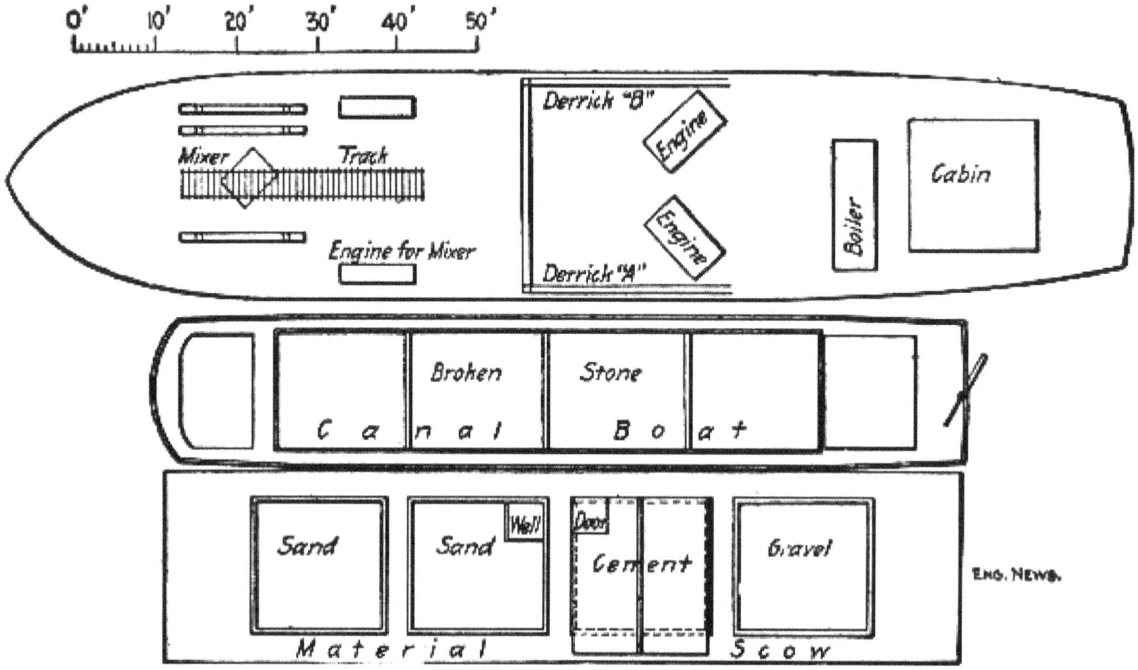

Fig. 80.—Sketch Plan of Concrete Mixing Plant for Buffalo Breakwater.

Figure 80 shows the arrangement of the mixing plant. The mixer was a 5-ft. cube mixer holding 125 cu. ft., mounted on a trestle and operated by a 9×12-in. horizontal engine taking steam from a 4×10-ft. locomotive boiler, also supplying steam to two derrick engines. The material scow contained two pockets for sand, one for gravel and one housed over for cement. Two inside cement men passed out the bags in lots of six to one outside cement man who cut and emptied them into the charging bucket. Three sand shovelers each loaded a 3.6 cu. ft. barrow and wheeled them tandem to the bucket, and two gravel men each loaded a 2.7 cu. ft. barrow and wheeled them tandem to the bucket. The broken stone was loaded by eight shovelers into another bucket, also containing 21.6 cu. ft. The two buckets were alternately hoisted and emptied into the mixer hopper, there being a dump man on the mixer who dumped the buckets and attended to the water supply. A charger put the mixer in operation and when the charge was mixed the car men dumped it into a skip resting on a small car which was then run out on the track under the mixer to the derrick which handled the skip to the work. Derrick *A* handled the materials from the scows and derrick *B* handled the mixed concrete. The force on the derricks consisted of two enginemen, four tagmen and the fireman.

The ten parapet wall sections containing 841 cu. yds. were built in 46 hours, making 17 batches of 1.07 cu. yds., or 18.2 cu. yds. placed per hour. The 17 parapet deck sections containing 1,720 cu. yds. were built in 88 hours, making 18.8 batches of 1.08 cu. yds., or 19.5 cu. yds. placed per hour. For the parapet deck work the force was increased by 2 men handling materials and 1 man on the mixer. The labor cost of mixing and placing the concrete was as follows:

Loading Gang—	Per day.	Per cu. yd.
1 assistant foreman	2.00	$0.011
3 cement handlers	5.25	0.029
3 sand shovelers	5.25	0.029
2 gravel shovelers	3.50	0.020
8 stone shovelers	14.00	0.076
1 hooker on	1.75	0.010
Totals	$31.75	$0.175
Mixer Gang—		
1 dumpman	$ 1.75	$0.010
1 charging man	1.75	0.010
2 car men	3.50	0.020
2 enginemen at $3.25	6.50	0.035
4 tagmen at $2	8.00	0.044
1 fireman	2.00	0.011
Totals	$23.50	$0.130
Wall Gang—		
1 Signalman	$ 1.75	$0.010
1 dumper	1.75	0.010
6 shovelers at $2	12.00	0.065
4 rammers	7.00	0.038
1 foreman	4.00	0.022

Totals	$26.50	$0.145
Grand totals	$81.75	$0.450

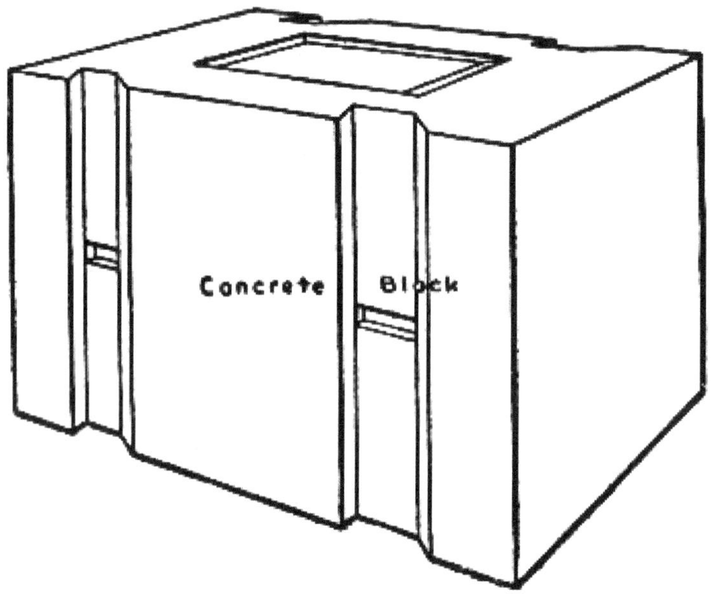

Fig. 81.—Concrete Blocks for Pier at Port Colborne Harbor.

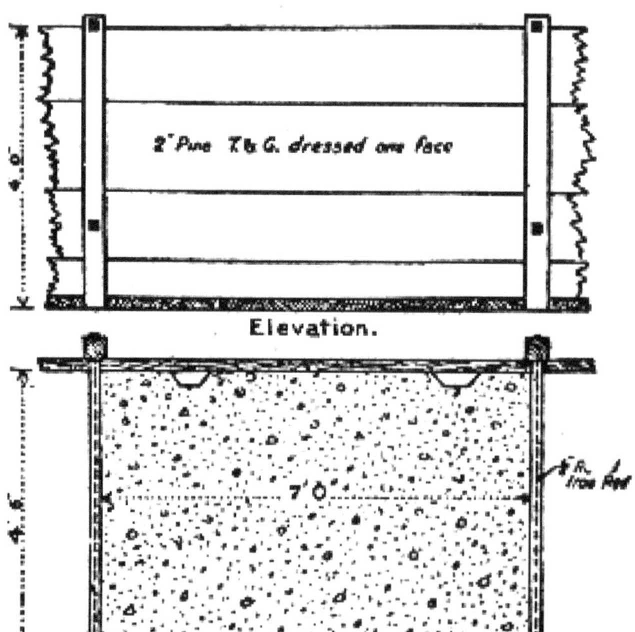

Fig. 82.—Forms for Molding Blocks, Port Colborne Harbor Pier.

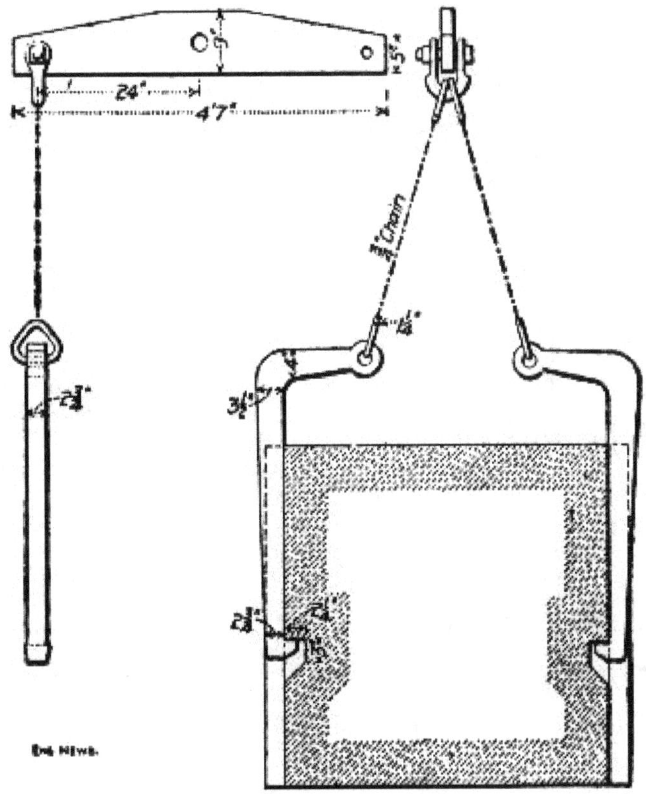

Fig. 83.—Device for Handling Blocks, Port Colborne Harbor Pier.

PIER CONSTRUCTION, PORT COLBORNE, ONT.—In constructing the new harbor at Port Colborne, Ont., on Lake Erie, the piers consisted of parallel rows of timber cribs set the width of the pier apart and filled in and between with stone blasted and dredged from the lake bottom in deepening the harbor. The tops of the cribs terminated below water level and were surmounted by concrete walls set on the outer edges. These walls were filled between with stone and the top of the filling was floored part way or entirely across, as the case might be, with a thick concrete slab. The footings of the walls to just above the water level were made of concrete blocks 4½×4×7 ft., constructed as shown by Fig. 81. The wall above the footing course and the floor slab were of concrete molded in place. The concrete work consisted of molding and setting concrete blocks and of molding concrete wall and slab in place.

The blocks were molded on shore, shipped to the work on scows and set in place by a derrick. Figure 82 shows the construction of the forms for molding the blocks; the bottom tie rods passed through the partitions forming the ends of the molds. The sides were removed in 48 hours and used over again. Figure 83 shows the hooks used for handling the molded blocks. Considerable trouble was had in setting these blocks level and close jointed, owing to the difficulty of leveling up the stone filling under water.

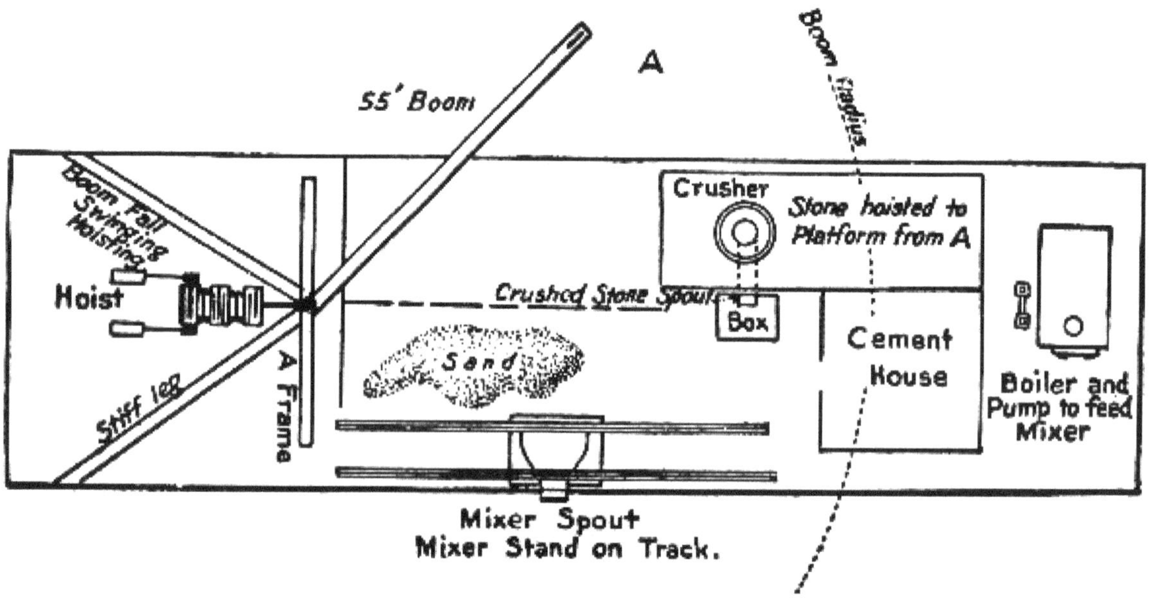

Fig. 84.—Scow Plant for Mixing and Placing Concrete, Port Colborne Harbor Pier.

The mass concrete was mixed and placed by the scow plant, shown by Fig. 84. The scow was loaded with sufficient sand and cement for a day's work and towed to and moored alongside the pier. Forms were set for the wall on top of the block footing. These forms were placed in lengths of 60 to 75 ft. of wall and resembled the block forms with partitions omitted. The bottoms of the rear uprights were held by being wedged into the grooves in the blocks, and the bottoms of the front uprights were held by bolts resting on top of the blocks. The tops of the uprights were held together across the wall by tie bolts. The forms being placed, the mode of procedure was as follows:

The crusher fed directly into a measuring box. After some 6 ins. of stone had run into the box the door of the crusher spout was closed. A wheelbarrow load of sand was spread over the stone in the box and over this were emptied and spread two or three bags of cement. Another layer of stone and then of sand and of cement were put in and these operations repeated until the box was full. The box was then hoisted and dumped into the hopper of a gravity mixer of the trough type which ran along a track on the scow and fed directly into the forms. The gang worked consisted of 1 foreman, 1 derrickman and 18 common laborers. This gang placed from 65 to 75 cu. yds. of concrete per day at a labor cost of 50 cts. per cu. yd.

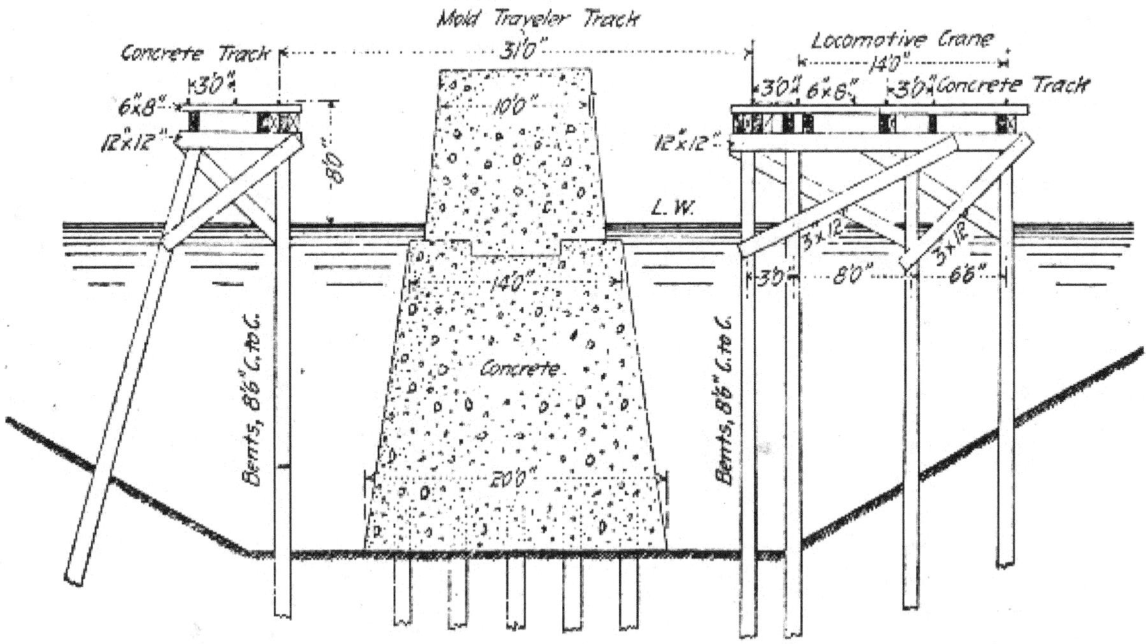

Fig. 85.—Cross-Section of Concrete Pier, Superior, Wis.

CONCRETE BLOCK PIER, SUPERIOR ENTRY, WIS.—The methods and cost of constructing a concrete pier 3,023 ft. long and of the cross-section shown by Fig. 85 at Superior entry, Wisconsin, are given in the following paragraphs.

Molds and Molding.—About 80 per cent. of the concrete was deposited in molds under water, according to a plan devised by Major D. D. Galliard, corps of engineers. In brief the concrete was built in place in two tiers of blocks, the lower tier resting directly on piles and being entirely under water and the upper tier being almost entirely above water. As shown by Fig. 85, a pile trestle was built on each side of the proposed pier and a traveler for raising and lowering the molds spanned the space between trestles.

The molds were bottomless boxes built in four pieces, two sides and two ends, held together by tie rods. Fig. 86 shows an end and a side of one of the shallow water molds and Fig. 87 shows in detail the method of fastening the end to the side. It will be seen that the 1¼-in. turnbuckle rods pass through the ends of beams that bear against the outside of the mold. These tie rods have eyes at each end in which rods with wedge-shaped ends are inserted. The molds were erected on the trestle by a locomotive crane and were then lifted by the mold traveler, carried and lowered into place. The largest one of these molds with its iron ballast, weighed 40 tons. To remove a mold, after the block had hardened, the nuts on the wedge-ended rods were turned, thus pulling the wedge end from the eye of the tie rod and releasing the sides of the mold from the ends.

The locomotive crane then raised the ends and sides, one at a time, and assembled them ready to be lowered again for the next block. The time required to remove one of these 40-ton molds, reassemble and set it again rarely exceeded 60 minutes and was sometimes reduced to 45 minutes.

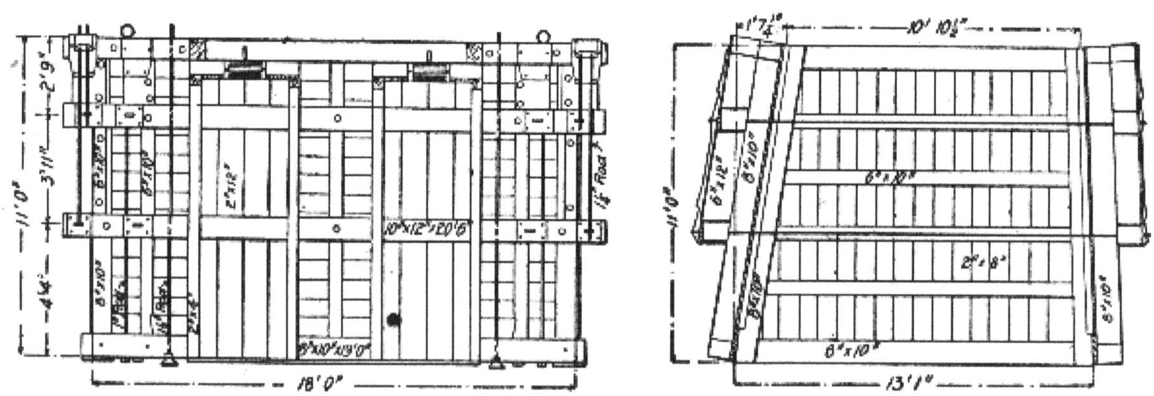

Fig. 86.—Mold for Concrete Block for Pier at Superior, Wis.

The concrete was deposited in alternate blocks and the molds described were for the first blocks; for the intermediate blocks molds of two side pieces alone were used, the blocks already in place serving in lieu of end pieces.

The two side pieces were bolted together with three tie rods at each end; the tie rods were encased in a box of 1-in. boards 4×4 ins. inside which served as a strut to prevent the sides from closing together and as a means of permitting the tie rods to be removed after the concrete had set. The mold was knocked down just as was the full mold described above and the boxes encasing the tie rods were left in the concrete.

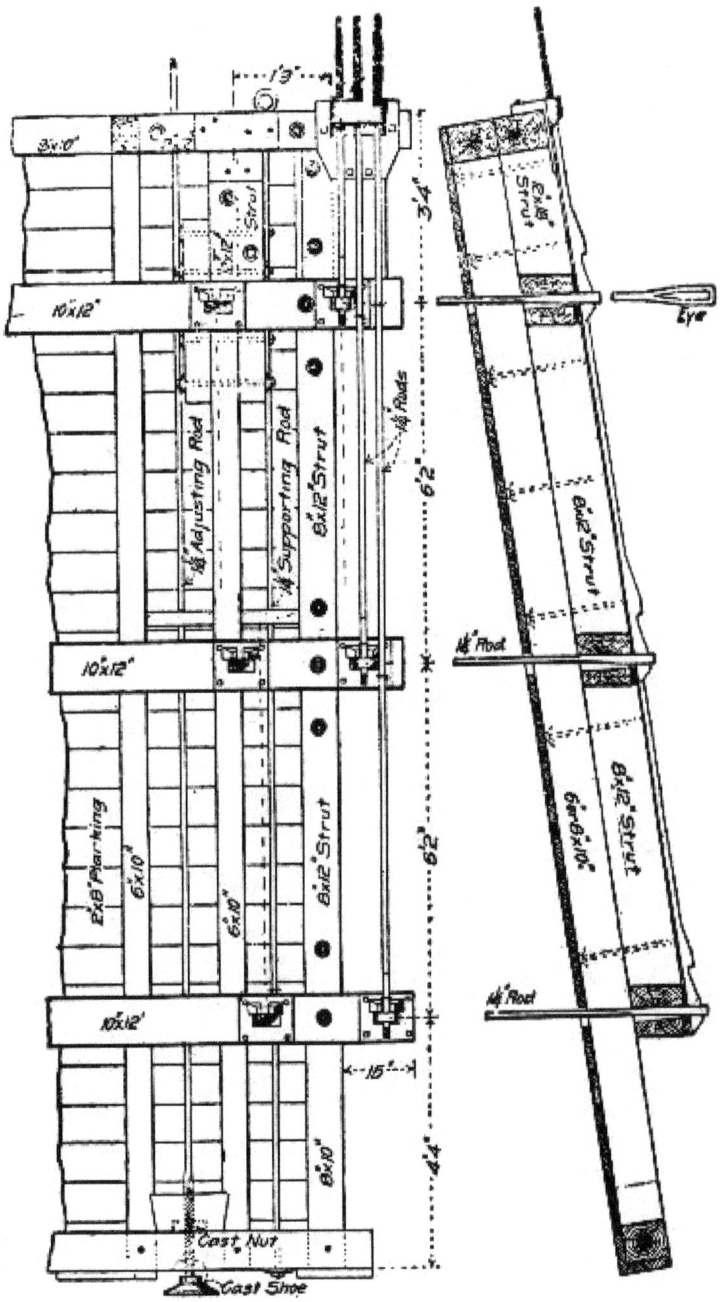

Fig. 87.—Device for Locking End and Side of Mold for Concrete Blocks for Pier at Superior, Wis.

An important feature was the device for handling the molds; this, as before stated, was a traveler, which straddled the pier site, it having a gage of 31 ft. It carried a four-drum engine, the drums of which were actuated, either separately or together, by a worm gear so as to operate positively in lowering as well as in raising. The load was hung from four hooks, depending by double blocks and ⅝-in. wire rope from four trolleys suspended from the trusses of the traveler; this arrangement allowed a lateral adjustment of the mold. The hoisting speed was 6 ft. per minute and the traveling speed 100 ft. per minute. The locomotive crane also deserves mention because it was mounted on a gantry high enough to permit material cars to pass under it on the same trestle, thus making it practicable to work two cranes.

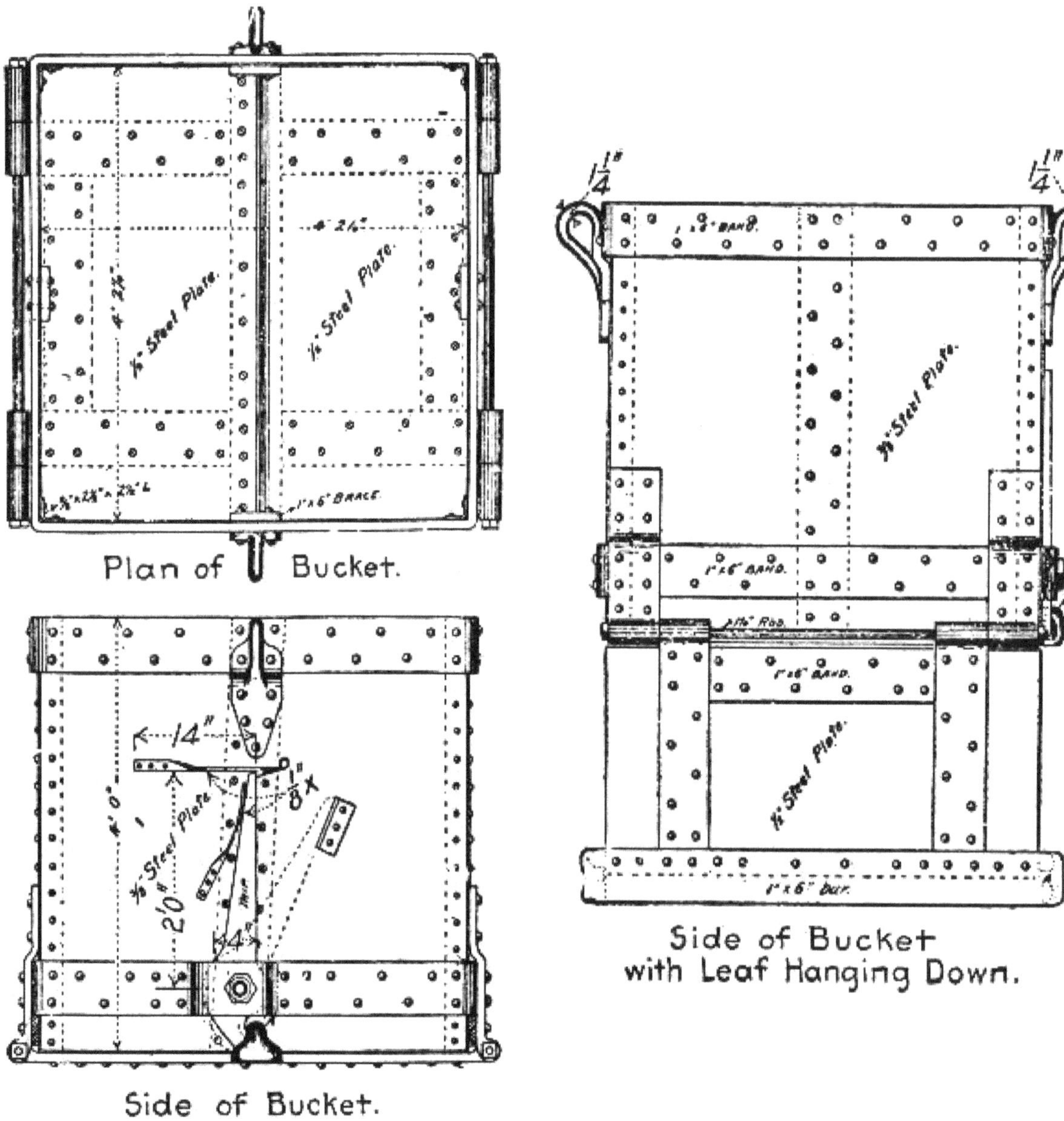

Fig. 88.—Bucket for Depositing Concrete Under Water for Pier at Superior, Wis.

The concrete was received from the mixer into drop bottom buckets of the form shown by Fig. 88. The buckets were taken to the work four at once on cars, and there lifted by the locomotive crane and lowered into the mold where they were dumped by tripping a latch connected by rope to the crane. To prevent the concrete from washing, the open tops of the buckets were covered with 3×4 ft. pieces of 12-oz. canvas in which were quilted 110 pieces of 1/16×1×3-in sheets of lead. Two covers were used on each bucket and were attached one to each side of the bucket top so as to fold over the top with a lap. This arrangement was entirely successful for its purpose.

Concrete Mixing.—The proportions of the subaqueous concrete were 1-2½-5 by volume, or 1-2.73-5.78 by weight, cement being assumed to weigh 100 lbs. per cu. ft.; the proportions of the superaqueous concrete were 1-3.12-6.25 by volume, or 1-3.41-7.22 by weight. The dry sand weighed 109.2 lbs. per cu. ft., the voids being 35.1 per cent.; the pebbles weighed 115.5 lbs. per cu. ft., the voids being 31 per cent.

The pebbles for the concrete were delivered by contract and were unloaded from scows by clam-shell bucket into a hopper. This hopper fed onto an endless belt conveyor which delivered the pebbles to a rotary screen. Inside this screen water was discharged under a pressure of 60 lbs. per sq. in. from a 4-in. pipe to wash the pebbles. From the screen the pebbles passed through a chute into 4-cu. yd cars which were hauled up an incline to a height of 65 ft. by means of a hoisting engine. The cars were dumped automatically, forming a stock pile. Under the stock pile was a double gallery or tunnel provided with eight chutes through the roof and from these chutes the cars were loaded and hauled by a hoisting engine up an inclined trestle to the bins above the concrete mixer. The sand was handled from the stock pile in the same manner. The cement was loaded in bags on a car in the warehouse, hauled to the mixer and elevated by a sprocket chain elevator.

Chutes from the bins delivered the materials into the concrete mixer, which was of the Chicago Improved Cube type, revolving on trunnions about an axial line through diagonal corners of the cube. The mixer possessed the advantage of charging and discharging without stopping. It was driven by a 7×10-in. vertical engine with boiler. The mixer demonstrated its ability to turn out a batch of perfectly mixed concrete every 1⅓ minutes. It discharged into a hopper provided with a cut-off chute which discharged into the concrete buckets on the cars.

Labor Force and Costs.—In the operation of the plant 55 men were employed, 43 being engaged on actual concrete work and 12 building molds and appliances for future work. The work was done by day labor for the government and the cost of operation was as follows for one typical week, when in six days of eight hours each, the output was 1,383 cu. yds., or an average of 230 cu. yds. per day. The output on one day was considerably below the average on account of an accident to the plant, but this may be considered as typical.

Pebbles from Stock Pile to Mixer—	Per cu. yd.
4 laborers at $2	$0.0348
1 engineman at $3	0.0131
Coal, oil and waste at $1.03	0.0043
Sand from Stock Pile to Mixer—	
5 laborers at $2	$0.0434
1 engineman at $2.50	0.0109
Coal, oil and waste at $0.82	0.0035
Cement from Warehouse to Mixer—	
5 laborers at $2	$0.0434
Mixing Concrete—	
1 engineman at $2.50	$0.0109
1 mechanic at $2.50	0.0108
Coal, oil and waste at $1.29	0.0056
Transporting Concrete—	
4 laborers at $2	$0.0348
1 engineman at $3	0.0130
Coal, oil and waste at $0.66	0.0028
Depositing Concrete in Molds—	
4 laborers at $2	$0.0348
1 engineman at $3	0.0130
1 rigger at $3	0.0130
Coal, oil and waste at $1.18	0.0051
Assembling, Transporting, Setting and Removing Molds—	

4 laborers at $2	$0.0347
1 engineman at $3.25	0.0141
1 carpenter at $3	0.0130
1 mechanic at $2.50	0.0109
Coal, oil and waste at $1.39	0.0060
Care of Tracks—	
1 laborer at $2	$0.0086
1 mechanic at $2.50	0.0109
Supplying Coal—	
3 laborers at $2	$0.0260
Blacksmith Work—	
1 laborer at $2	$0.0086
1 blacksmith at $3.25	0.0141
1 waterboy at $0.75	0.0032
Total per cubic yard	$0.4473
Add 75% of cost of administration	0.1388
Total labor per cu. yd.	$0.5861

The total cost of each cubic yard of concrete in place was estimated to be as follows:

	Per cu. yd.
Ten-elevenths cu. yd. pebbles at $1.085	$0.9864
Ten-twenty seconds cu. yd. sand at $0.00	0.0000
1 26 bbls. cement at $1.77	2.2302
Labor as above given	0.5861
Cost of plant distributed over total yardage	0.8400
Total	$4.6427

It will be noted that the sand cost nothing as it was dredged from the trench in which the pier was built, and paid for as dredging. The cost of the plant is distributed over this south pier and over the proposed north pier work on the basis of only 20 per cent. salvage value after the completion of both piers. It is said, however, that 80 per cent. is too high an allowance for the probable depreciation.

DAM, RICHMOND, INDIANA.—The dam shown in cross-section in Fig. 89 was built at Richmond, Ind. It was 120 ft. long and was built between the abutments of a dismantled bridge. The concrete was made in the proportion of 1 bbl. Portland cement to 1 cu. yd. of gravel; old iron was used for reinforcement. The foundations were put down by means of a cofferdam which was kept dry by pumping. On completion it was found that there was a tendency to scour in front of the apron and accordingly piling was driven and the intervening space rip-rapped with large stone. Labor was paid as follows per day: Foreman, $3; carpenter, $2.50; cement finisher, $2; laborers, $1.50. The concrete was mixed by hand and wheeled to place in wheelbarrows. The cost of the work was as follows:

Materials—	Per cu. yd.
204 bbls. cement at $1.60	$1.485
Sand and gravel	0.800
Lumber	0.610
Tools, hardware, etc.	0.445
Total materials	$3.34
Labor—	
Clearing and excavating	$0.96
Setting forms and mixing concrete	1.01
Pumping	0.27
Total labor	$2.24
Total materials and labor	$5.58

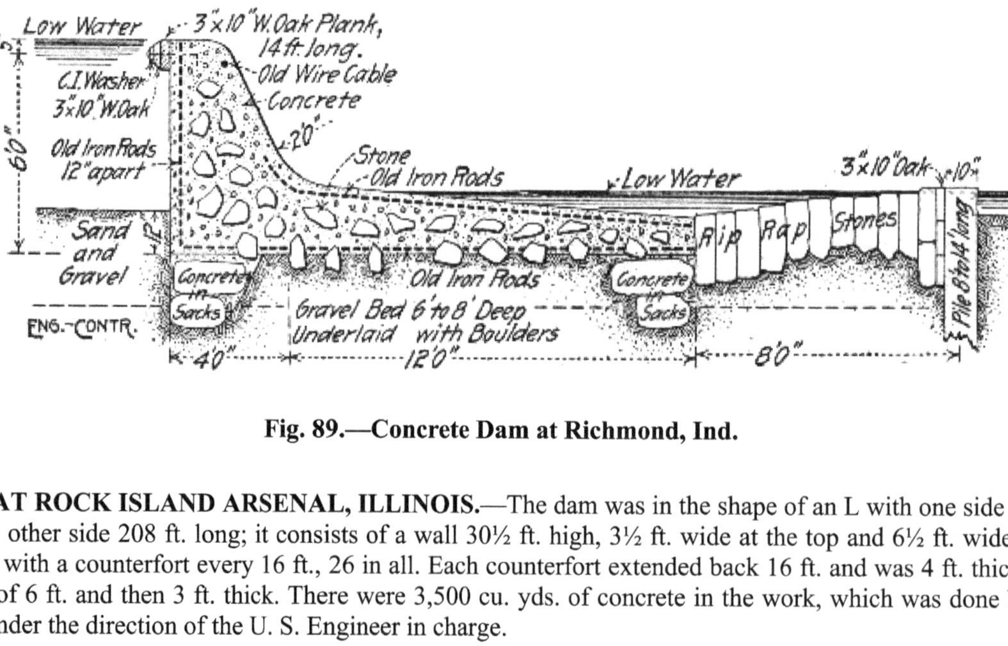

Fig. 89.—Concrete Dam at Richmond, Ind.

DAM AT ROCK ISLAND ARSENAL, ILLINOIS.—The dam was in the shape of an L with one side 192 ft. and the other side 208 ft. long; it consists of a wall 30½ ft. high, 3½ ft. wide at the top and 6½ ft. wide at the bottom with a counterfort every 16 ft., 26 in all. Each counterfort extended back 16 ft. and was 4 ft. thick for a height of 6 ft. and then 3 ft. thick. There were 3,500 cu. yds. of concrete in the work, which was done by day labor under the direction of the U. S. Engineer in charge.

The forms consisted of front and back uprights of 8×10-in. stuff 24 ft. high, connected through the wall by ¾-in. rods which were left in the concrete. The lagging was 2×12-in. plank dressed down 1¾ ins. placed inside the uprights. These forms were built full height in 16-ft. sections with a counterfort coming at the center of each section. Each section contained 95 cu. yds. of concrete and was filled in a day's work. The concrete was a 1-4-7 mixture wet enough to quake when rammed. Run of crusher limestone was used of which 50 per cent. passed a 1-in. sieve, 17 per cent. a No. 3 sieve and 9 per cent. a No. 8 sieve. The concrete was mixed in Cockburn Barrow & Machine Co.'s screw-feed mixer which discharged into 2-in. plank skips 2 ft. wide 5⅓ ft. long and 14 ins. deep, holding ¼ cu. yd. These skips were taken on cars to a derrick crane overhanging the forms and by it hoisted and dumped into the forms. The derrick was moved along a track at the foot of the wall as the work progressed. The concrete was spread and rammed in 6-in. layers. The men were paid $1.50 per 8-hour's work and the work cost including footing, as follows:

Item—	Total.	Per cu. yd.
Cement	$1,500.00	$0.429
Sand	400.00	0.114
Storing and hauling cement	460.00	0.131
Taking sand from barge to mixer	96.00	0.027
Crushing stone	1,450.00	0.414
Mixing concrete	4,825.00	1.378
Placing concrete	1,670.00	0.477
Lumber for forms, etc.	600.00	0.171
Erecting and taking down forms	2,450.00	0.700
Totals	$13,451.00	$3.841

DAM AT McCALL FERRY, PA.—The dam was 2,700 ft. long and 48 ft. high of the cross-section shown by Fig. 90 and with its subsidiary works required some 350,000 cu. yds. of concrete. The plant for mixing and placing the concrete was notable chiefly for its size and cost. Parallel to the dam, which extended straight across the river, and just below its toe a service bridge consisting of a series of 40-ft. concrete arch spans was built across the river. This service bridge was 50 ft. wide and carried four standard gage railway tracks besides a traveling crane track of 44 ft. gage. This very heavy construction of a temporary structure was necessitated by the frequency of floods against which only a solid bridge could stand; it was considered cheaper in the long run to provide a bridge which would certainly last through the work than to chance a structure of less cost which would certainly go out with the floods. The concrete service bridge was designed to be destroyed by blasting when the dam had been completed. The method of construction was to build the dam in alternate 40 ft. sections, mixing the concrete on shore, taking it out along the service bridge in buckets on cars and handling the buckets from cars to forms by traveling cranes.

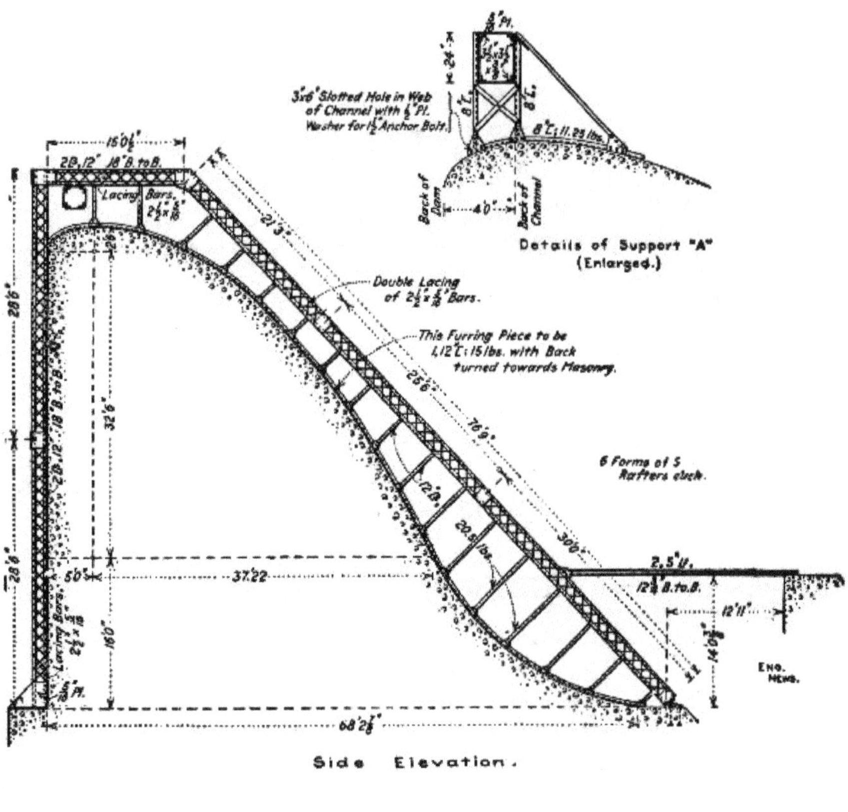

Fig. 90.—Steel Forms for McCall Ferry Dam.

The concrete mixing plant is shown by Fig. 91. Cars loaded with cement, sand and stone were brought in over the tracks carried on the wall tops of the bins and were unloaded respectively into bins *A*, *B* and *C*, of which there were eight sets. Each set supplied material by means of measuring cars to a 1 cu. yd. Smith mixer. Two sets of cars were used for each mixer so that one could be loading while the other was charging. The mixers discharged into 1 cu. yd. buckets set two on a car and eight cars were hauled to the work in train by an 18-ton "dinky." At the work the buckets were picked up by the traveling cranes and the concrete dumped into the forms. Figure 90 shows the construction of the steel forms. Six sets of forms were used. Each set consisted of five frames spaced 10 ft. apart and braced together in the planes parallel to the dam, and each set molded 40 ft. of dam. The lagging consisted of wooden boxes 8½ ft. wide and 10 ft. long. For the vertical face of the dam these boxes were attached by bolts to the vertical post, for the curved face they were bolted to a channel bent to the curve and held by struts from the inclined post of the steel frame.

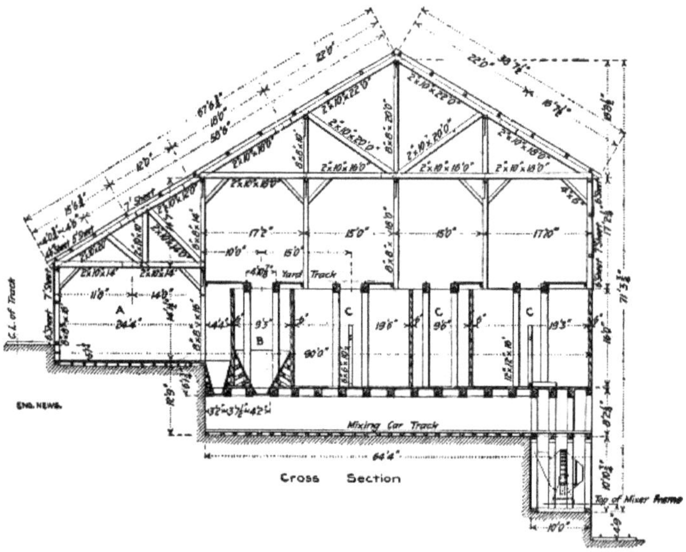

Fig. 91.—Concrete Mixing Plant for McCall Ferry Dam.

In construction the footing and the body of the dam to an elevation of 5 ft. above the beginning of the curve were built continuously across the river; above this elevation the dam was built in alternate 40-ft. sections. The strut back to the service bridge shown in the lower right hand corner of Fig. 90, shows the manner of bracing the first 30-ft. section of the inclined post to hold the lagging for the continuous portion. The lagging was added a piece at a time as concreting progressed. The ends of each set of frames for a 40-ft. section were for the isolated sections closed by timber bulkheads carrying box forms to mold grooves into which the concrete of the intermediate sections would bond.

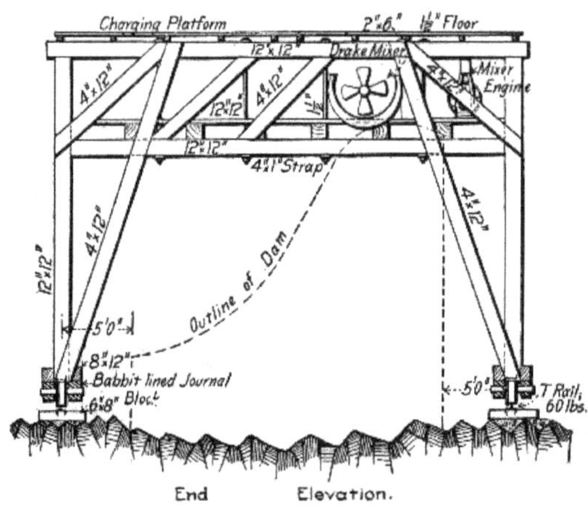

Fig. 92.—Traveler for Concreting Dam, Chaudiere Falls, Quebec.

The concrete used was a 1-3-5 mixture, the stone ranging in size from 2 to 5 ins. Rubble stone from one man size to ½ ton were bedded in the concrete. The capacity of the concrete plant was 2,000 cu. yds. per day or about 250 cu. yds. per mixer per 10-hour day.

DAM, CHAUDIERE FALLS, QUEBEC.—The dam was 800 ft. long and from 16 to 20 ft. high, constructed of 1-2-4 concrete with rubble stone embedded. The rubble stones were separated at least 9 ins. horizontally and 12 ins. vertically and were kept 20 ins. from faces. At one point the rubble amounted to 40 per cent. of the volume, but the average for the dam was 25 to 30 per cent. The stone was broken at the work, some by hand, but most by machine, all to pass a 2-in. ring. Hand-broken stone ran very uniform in size and high in voids, often up to 50 per cent. Stone broken by crusher with jaws 2 ins. apart would run 20 to 30 per cent. over 2 ins. in size and give about 45 per cent. voids; with crusher jaws 1½ ins. apart from 98 to 100 per cent. was under 2 ins. in size and contained about 42 per cent. of voids. It was found that if the crushers were kept full all the time the product was much smaller, particularly with Gates gyratory crusher, though a little more than rated power was required when the crusher was thus kept full. This practice secured increased economy in both quantity and quality of product. The concrete was made and placed by means of a movable traveler shown by Fig. 92. Concrete materials were supplied to the charging platform of the traveler by means of a traveling derrick moving on a parallel track. In placing the concrete on the rock bottom it was found necessary in order to secure good bond to scrub the rock with water and brooms and cover it with a bed of 2 ins. of 1-2 mortar. The method of concreting in freezing weather is described in Chapter VII.

12

BRIDGE PIERS AND ABUTMENTS CONSTRUCTION

The construction of piers and abutments for bridges is best explained by describing individual examples of such work. So far, in America, bridge piers have been nearly always of plain concrete and of form and section differing little from masonry piers; where reinforcement has been used at all it has consisted of a surface network of bars introduced chiefly to ensure monolithic action of the pier under lateral stresses. In Europe cellular piers of reinforced concrete have been much used. Plain concrete abutments differ little in form and volume from masonry abutments. Reinforced concrete abutments are usually of L-section with counterforts bracing the upright slab and bridge seat to the base slab.

Form work for reinforced abutments is somewhat complex; that for plain abutments and piers is of simple character, the only variations from plain stud and sheathing construction being in the forms for moldings and coping and for cut-waters. For piers of moderate height the form is commonly framed complete for the whole pier, but for high piers it is built up as the work progresses by removing the bottom boards and placing them at the top. Opposite forms are held together by wire ties through the concrete. Movable panel forms have been successfully employed, but they rarely cheapen the cost much. Sectional forms, which can be shifted from pier to pier where a number of piers of identical size are to be built, may frequently be used to advantage. An example of such use is given in this chapter.

Derricks are the recognized appliances for hoisting and placing the concrete in pier work; they are the only practicable appliance where the pier is high and particularly where it stands in water and mixing barges are employed. For abutment work and land piers of moderate height derricks and wheelbarrow or cart inclines are both available and where much shifting of the derricks is involved the apparently more crude method compares favorably in cost.

The methods of placing concrete under water for pier foundations are described in Chapter V, and the use of rubble concrete for pier construction is illustrated by several examples in Chapter VI. The following examples of pier and abutment construction cover both large and small work and give a clear idea of current practice.

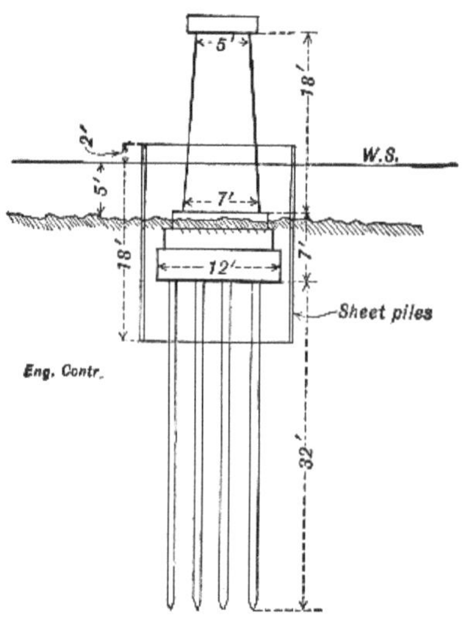

Fig. 93.—Pier and Cofferdam for a Railway Bridge.

COST OF CONSTRUCTING RECTANGULAR PIER FOR A RAILWAY BRIDGE.—This pier, Fig. 93, was built in water averaging 5 ft. deep. The cofferdam consisted of triple-lap sheet piling, of the Wakefield pattern, the planks being 2 ins. thick, and spiked together so as to give a cofferdam wall 6 ins thick. The cofferdam enclosed an area 14×20 ft., giving a clearance of 1 ft. all around the base of the concrete pier, and a clearance of 2 ft. between the cofferdam and the outer edge of the nearest pile. The cofferdam sheet piles were 18 ft. long, driven 11 ft. deep into sand, and projecting 2 ft. above the surface of the water.

The concrete base resting on the foundation piles was 12×18 ft. The concrete pier resting on this base was 7×13 ft. at the bottom, and 5×11 ft. at the top. The pier supported deck plate girders. There were 100 cu. yds. of concrete in the pier and base.

The cost of this pier, which is typical of a large class of concrete pier work, has been obtained in such detail that we analyze it in detail, giving the costs of cofferdam construction and excavation as well as of mixing and placing the concrete.

Setting up and taking down derrick and platform:	
4 days foreman at $5.00	$ 20.00
¾ days engineman at $3.00	2.25
¾ days blacksmith at $3.00	2.25
¾ days blacksmith helper at $2.00	1.50
22 days laborers at $2.00	44.00
Total	$ 70.00
Cofferdam—	
7 days foreman at $5.00	$ 35.00
4 days engineman at $3.00	12.00
38 days laborers at $2.00	76.00
1 ton coal at $3.00	3.00
Total labor on 7,900 ft. B. M. at $16.00	$126.00
7,900 ft. B. M. at $20.00	158.00
Total for 58 cu. yds. excavation	$284.00
Wet Excavation—	
1.8 days foreman at $5.00	$ 9.00
1.5 days engineman at $3.00	4.50
9 days laborers at $2.00	18.00
½ ton coal at $3.00	1.50
Total labor on 58 cu. yds. at 57c.	$ 33.00
Foundation Piles—	
960 lin. ft. at 10c	$ 96.00
4 days setting up driver and driving 24 piles at $20 per day for labor and fuel	80.00
Total	$176.00
Concrete—	
100 cu yds. stone at $1.00	$100.00
40 cu. yds. sand at $0.50	20.00
100 bbls. cement at $2.00	200.00
5 days foreman at $5.00	25.00
50 days laborers at $2.00	100.00
5 days engineman at $3.00	15.00
2 tons coal at $3.00	6.00

Total, 100 cu. yds. at $4.66	$466.00
8 days carpenters at $3.00	24.00
2,400 ft. B. M. 2-in. plank at $25.00	60.00
1,000 ft. B. M. 4×6-in. studs at $20.00	20.00
Nails, wire, etc	2.00
Total forms for 100 cu. yds. at $1.06	$106.00
Summary—	
Setting up derrick, etc.	$ 70.00
Cofferdam (7,900 ft. B. M.)	284.00
Wet excavation (58 cu. yds.)	33.00
Foundation piles (24)	176.00
Concrete (100 cu. yds.)	466.00
Forms (3,400 ft. B. M.)	106.00
Total	$1,135.00
Transporting plant	20.00
20 days rental of plant at $5.00	100.00
Total cost of pier	$1,252.00

Regarding the item of plant rental, it should be said that the plant consisted of a pile driver, a derrick, a hoisting engine, and sundry timbers for platforms. There was no concrete mixer. Hence an allowance $5 per day for use of plant is sufficient.

It will be noted that no salvage has been allowed on the lumber for forms. As a matter of fact, all this lumber was recovered, and was used again in similar work.

Referring to the cost of cofferdam work, we see that, in order to excavate the 58 cu. yds. inside the cofferdam, it was necessary to spend $284, or nearly $5 per cu. yd. before the actual excavation was begun. The work of excavating cost only 57 cts. per cu. yd., but this does not include the cost of erecting the derrick which was used in raising the loaded buckets of earth, as well as in subsequently placing the concrete. The sheet piles were not pulled, in this instance, but a contractor who understands the art of pile pulling would certainly not leave the piles in the ground. A hand pump served to keep the cofferdam dry enough for excavating; but in more open material a power pump is usually required.

The above costs are the actual costs, and do not include the contractor's profits. His bid on the work was as follows:

Piles delivered	12 cts. per ft.
Piles driven	$5 each
Cofferdam	$37 per M.
Wet excavation	$1.00 per cu. yd.
Concrete	$8.00 per cu. yd.

In order to ascertain whether or not these prices yielded a fair profit, it is necessary to distribute the cost of the plant transportation and rental over the various items. We have allowed $120 for plant transportation and rental, and $70 for setting up and taking down the plant, or $190 in all. The working time of the plant was as follows:

	Days.	Per cent. of time.	Prorated plant cost.
Cofferdam	7	39	$74
Excavation	2	11	21
Foundation piles	4	22	42

Concrete	5	28	53
Totals	18	100	$190

As above given, the labor on the 7,900 ft. B. M. in the cofferdam cost $126, or $16 per M.; but this additional $74 of prorated plant costs, adds another $9 per M., bringing the total labor and plant to $25 per M., to which must be added the $20 per M. paid for the timber in the cofferdam, making a grand total of $45 per M. This shows that the contractor's bid of $37 per M. was much too low.

The labor on the excavation cost 57 cts. per cu. yd., to which must be added the prorated plant cost of $21 distributed over the 58 cu. yds., or 36 cts. per cu. yd., making a total of 93 cts. per cu. yd. This shows that the bid of $1 per cu. yd. was hardly high enough.

The labor on the 24 foundation piles cost $80, or $3.33 each. The prorated plant cost is $42, or $1.75 per pile, which, added to $3.33, makes a total of $5.08. This shows that the bid of $5 Per pile for driving was too low. However there was a profit of 2 cts. per ft., or 80 cts. per pile, on the cost of piles delivered.

The concrete amounted to 100 cu. yds. Hence the prorated plant cost of $53 is equivalent to 53 cts. per cu. yd. Hence the total cost of the concrete was:

	Per cu. yd.
Cement, sand and stone	$3.20
Foreman (at $5)	0.25
Labor (at $2)	1.00
Engineman (at $3)	0.15
Coal (at $3)	0.06
Carpenters (at $3)	0.24
Forms (at $23.50, used once)	0.80
Wire, nails, etc	0.02
Prorated plant cost	0.53
Total	$6.25

Since the contract price for concrete was $8 per cu. yd., there was a good profit in this item.

BACKING FOR BRIDGE PIERS AND ABUTMENTS.—Six piers and two abutments of the City Island bridge were constructed in 1906 at New York city, of masonry backed with 1-2-4 concrete below and 1-3-5 concrete above high water. The piers and abutments were all sunk to rock or hard material by means of timber cofferdams. Table XVI gives the labor cost of mixing and placing the concrete backing for one abutment and three piers, after the materials were delivered on the scows. The concrete was mixed by a rectangular horizontal machine mixer and deposited by 2-cu. yd. bottom dump buckets handled by derrick scows and stiff leg derricks. The high cost of concreting on Pier 2 was due to the fact that the concrete was improperly deposited and had to be removed and the higher cost in Abutment 1 was probably due to the fact that the abutment was so long and narrow that it was difficult to handle the bucket.

Table XVI.—Cost of Concrete Backing for Masonry Piers.

		Abutment No. 1.			Pier No. 2.			Pier No. 3.			Pier No. 4.			Summary.	
	Wages per Hour.	No. hrs.	Total Cost	Cost per cu. yd.	No. hrs.	Total Cost	Cost per cu. yd.	No. hrs.	Total Cost	Cost per cu. yd.	No. hrs.	Total Cost	Cost per cu. yd.	Total Hrs.	Total cost.
Superintendent	70	24	$16.80	$0.03	47	$32.90	$0.09	72	$50.40	$0.05	16	$11.20	$0.03	159	$111.30
Foreman	35	160	56.00	0.09	128	44.80	0.13	324	113.40	0.12	54	18.90	0.06	666	233.10

Laborers	15-20	2555	383.25	0.65	2038	313.60	0.92	3513	526.95	0.56	940	141.00	0.44	9046	1364.90
Engineman	30	365	109.50	0.19	196	58.50	0.19	244	73.20	0.08	60	18.00	0.06	865	259.50
Timekeeper	40	86	34.40	0.06	46	18.40	0.06	81	32.40	0.04	10	4.00	0.01	223	89.20

PNEUMATIC CAISSONS, WILLIAMSBURG BRIDGE.—Mr. Francis L. Pruyn, Assoc. M. Am. Soc. C. E., gives the following costs of concreting the pneumatic caissons for the Brooklyn tower of the Williamsburg bridge at New York city. The work comprised the mixing and placing of some 13,637 cu. yds. of concrete in two caissons. Table XVII shows the itemized costs for one caisson and Table XVIII shows them for the other caisson. The methods of work were as follows:

After each caisson was built it was towed to its proper site, where it was held in place by temporary pile dock built completely around it. On these docks the concrete was placed; a 2 cu. yd. cubical mixer of the usual pattern being used for mixing. The concrete materials, consisting of sand, stone and cement was handled direct from barges alongside, into the mixer. The concrete was placed by a derrick located in the center of the caisson, which was a bad feature as the caisson was usually out of level and considerable difficulty was experienced in swinging the derrick. On the South caisson ¾ cu. yd. bottom dump buckets were used in placing the concrete, on the North caisson the size of these was increased to 1½ cu. yd. which reduced the cost of placing 15 cts. per cu. yd. There were placed in the South caisson 3,827 cu. yds. in 32 days of actual working time—120 cu. yds. per day of 10 hrs. The gross time was 2 months. On the North caisson 5,693 cu. yds. were placed in 46 days worked—124 cu. yds. per day. The gross time was 4 months.

The rates of labor were as follows per 10-hour day:

Foreman	$5.00
Assistant foreman	2.50
Hoisters	2.50
Fireman	1.60
Laborer	1.50

Proportions concrete were 1: 2.5: 6.

The low price of sand in the North caisson was brought about by the finding of good building sand in the excavation for the anchorage, which work was done by the same contractor.

When the caissons had been sealed the iron material shafts were removed. This left holes 5 ft.×6 ft. extending from the roof of the caisson up to Mean H.W. which were filled with concrete. These shaft holes were 80 ft. deep on the South caisson and 100 ft. deep on the North caisson. They were partially filled with water and the concrete had to be placed with considerable care. Wooden chutes were used on the South caisson; they rested on the caisson roof, were filled with concrete and then raised allowing concrete to flow out at the bottom. The shaft holes were too deep on the North caisson for chutes and 20 cu. ft. bottom dump buckets were used. They had to be lowered to bottom of shaft each trip before dumping, a slow operation, which greatly added to the cost. Proportion for concrete 1-2.5-6.

The proportion for concrete in working chamber was the same as for all other concrete. The specifications called for 6 in. of mortar, of 1 part of cement to 2½ parts of sand, between the concrete and all bearing areas; that is, under the cutting edge and directly under the roof of the working chamber. The concrete was mixed in the cubical mixer and dumped on the bottom door of the material lock, the top door of the lock was then closed, the bottom door opened and the concrete fell through the shaft to the working chamber. It was then shoveled by the sand hogs into place. A 6-in. space was left below all bearing surfaces into which damp mortar was tightly rammed. Concreting the South caisson took 10¼ working days of 24 hours, the gangs working night and day in twelve 2-hour shifts; 1,566 cu. yds. of concrete and mortar were placed, or at the rate of 140 cu. yds. per 24 hours. The gross time including Sundays was 14½ days. The sand hogs worked in shifts of 2 hours each and received $3.50 for the two hours work. The twelve foremen received 1 dollar more: the average gang consisted of 12 sand hogs.

On the North caisson the organization was much better, owing to the experience gained on the first caisson; and in spite of the fact that the sand hogs, on account of the increased depth, received $4.00 for 1½ hours' work, or an increase of $22.00 per man per 24 hrs. over that on the South caisson, the work was done for less money. There were placed 1,566 cu. yds. of concrete in 7 working days of 24 hrs., or at the rate of 224 cu. yds. per day. The gross time was 11½ days including Sundays. The average number of men in the sand hog gangs was 18, with one foreman, who received $5 for 1½ hours work.

TABLE XVII.—ITEMIZED COST OF CONCRETING SOUTH CAISSON FOR BROOKLYN TOWER OF THE WILLIAMSBURG BRIDGE: COST OF CONCRETING CAISSONS ABOVE ROOF.

	South Caisson (3,827 cu. yds.).		
Materials.	Quantity.	Rate.	Amount.
Cement	4,480 bbls.	$1.57	$7,034.00
Sand	1,288 cu. yds.	.60	773.00
Broken stone	3,421 cu. yds.	1.50	5,132.00
Water			36.00
Total	3,827 cu. yds.	$3.39	$12,975.00
Labor.			
Mixing and placing	3,827 cu. yds.	$0.90	$3,432.00
Plant charges			2,280.00
Plant labor			742.00
Total plant	3,827 cu. yds.	$0.79	$3,022.00
Total cost	3,827 cu. yds.	$5.08	$19,429.00
General expenses, 10%	3,827 cu. yds.	.51	1,943.00
Grand total	3,827 cu. yds.	$5.59	$21,372.00

COST OF CONCRETING SHAFTS.

	South Caisson.		
Materials.	Quantity.	Rate.	Amount.
Concrete	612½ bbls.	$1.57	$962.00
Sand	193 cu. yds.	.40	77.00
Stone	493 cu. yds.	1.10	542.00
Total	541 cu. yds.	$2.92	$1,581.00
Labor.			
Handling, mixing and placing	541 cu. yds.	$0.96	$519.00
Plant charges, etc.	541 cu. yds.	1.06	576.00
Total	541 cu. yds.	$4.94	$2,676.00
General expenses, 10%	541 cu. yds.	.49	267.00
Grand total	541 cu. yds.	$5.43	$2,943.00

COST OF CONCRETE IN WORKING CHAMBERS.

	South Caisson. (1,435 cu. yds.)		
Materials.	Quantity.	Rate.	Amount.
Cement for concrete	1,666 bbls.	$1.57	$2,615.00
Cement for mortar	459 bbls.	1.57	720.00
Sand for both	670 cu. yds.	.40	268.00

Broken stone	1,181 cu. yds.	1.10	1,299.00
Total materials	1,435 cu. yds.	$3.42	$4,902.00
Labor.			
Top labor, mixing and placing	1,435 cu. yds.	$1.09	$1,575.00
Pneumatic labor	1,435 cu. yds.	4.93	7,117.00
Compressor house labor	1,435 cu. yds.	.19	275.00
Total labor	1,435 cu. yds.	$6.21	$8,967.00
Plant.			
Coal at $2.40 per ton	1,435 cu. yds.	.10	140.00
Concrete plant	1,435 cu. yds.	.79	1,145.00
Pneumatic plant	1.435 cu. yds.	1.05	1,522.00
Total plant	1,435 cu. yds.	$1.94	$2,807.00
Totals	1,435 cu. yds.	$11.57	$16,676.00
General expenses, 10%	1,435 cu. yds.	1.16	1,667.00
Grand total	1,435 cu. yds.	$12.73	$18,343.00

TABLE XVIII.—ITEMIZED COST OF CONCRETING NORTH CAISSON FOR BROOKLYN TOWER OF THE WILLIAMSBURG BRIDGE:

COST OF CONCRETING CAISSON ABOVE ROOF (5,692 cu. yds.)

Materials.	Quantity.	Rate.	Amount.
Cement	6,707½ bbls.	$1.57	$10,531.00
Sand	2,133 cu. yds.	.40	845.00
Broken stone	4,938 cu. yds.	1.10	5,432.00
Water			51.00
Total	5,692 cu. yds.	$2.96	$16,859.00
Labor.			
Mixing and placing	5,692 cu. yds.	$0.73	$4,159.00
Plant charges			2,952.00
Plant labor			517.00
Total	5,692 cu. yds.	$0.61	$3,469.00
Total cost	5,692 cu. yds.	$4.30	$24,487.00
General expenses, 10%	5,692 cu. yds.	.43	2,448.00
Grand total	5,692 cu. yds.	$4.73	$26,935.00

COST OF CONCRETING SHAFTS.

Materials.	Quantity.	Rate.	Amount.
Cement	614½ bbls.	$1.57	$965.00
Sand	204 cu. yds.	.40	82.00

Stone	521 cu. yds.	1.10	574.00
Total	576 cu. yds.	$2.82	$1,621.00
Labor			
Mixing and placing	576 cu. yds.	1.70	982.00
Plant charges, etc.	576 cu. yds.	1.36	795.00
Total	576 cu. yds.	$5.88	$3,398.00
General expenses, 10%	576 cu. yds.	.59	339.00
Grand total	576 cu. yds.	$6.47	$3,737.00

COST OF CONCRETING WORKING CHAMBERS (1,566 cu. yds.).

Materials.	Quantity.	Rate.	Amount.
Cement for concrete	1,559 bbls.	$1.51	$2,446.00
Cement for mortar	442 bbls.	1.51	$694.00
Sand for both	630 cu. yds.	.40	252.00
Broken stone	1,380 cu. yds.	1.10	1,518.00
Total	1.566 cu. yds.	$3.14	$4,910.00
Labor.			
Top labor, mixing and placing	1,566 cu. yds.	$0.78	$1,198.00
Pneumatic labor	1,566 cu. yds.	4.91	7,694.00
Compressor house labor	1,566 cu. yds.	.11	180.00
Total labor	1,566 cu. yds.	$5.80	$9,072.00
Plant.			
Coal at $2.40 per ton	1,566 cu. yds.	.06	87.00
Concrete plant	1,566 cu. yds.	.86	1,352.00
Pneumatic plant	1,566 cu. yds.	.81	1,272.00
Total plant	1,566 cu. yds.	$1.73	$2,711.00
Totals	1,566 cu. yds.	$10.67	$16,693.00
General expenses, 10%	1,566 cu. yds.	1.06	1,669.00
Grand total	1,566 cu. yds.	$11.73	$18,362.00

COST OF FILLING PIER CYLINDERS.—The following costs were obtained in mixing and placing concrete in steel cylinder piers. The sand and gravel were wheeled 100 ft. to the mixing board at the foot of the cylinder, mixed and shoveled into wooden skips, hoisted 20 ft. by horsepower and dumped into the cylinder. The foreman worked on the mixing board and the men worked with great energy. The costs were as follows:

Item—	Per day.	Per cu. yd.
6 men wheeling materials and mixing at 15 cts. per hour	9.00	$0.45
2 men dumping skips and ramming at 15 cts. per hour	3.00	0.15
1 team and driver at 40 cts. per hour	4.00	0.20
1 foreman at 30 cts. per hour	3.00	0.15
	——	——
Totals	$19.00	$0.95

PIERS, CALF KILLER RIVER BRIDGE.—The following methods and costs of building two new piers and extending three old piers with concrete are given by Mr. J. Guy Huff. The work was done by the railway company's masonry gangs. Figure 94 shows the arrangement of the several piers and the character of the work on each and Fig. 95 gives the detail dimensions of the three main piers.

The sand and aggregate, consisting of blast furnace slag, were unloaded from cars to platforms on a level with the top of rail, placed about 100 ft. south from the south end of the bridge. A cubical 1/6 cu. yd. mixer was used. This was operated by a gasoline engine, and was located on a platform about 50 ft. south of the south end pier. A tank near the mixer to supply water was elevated enough to get the desired head, and was kept filled by a pump run by another gasoline engine located down by the river bank. The cement house was located between the mixer platform and slag pile.

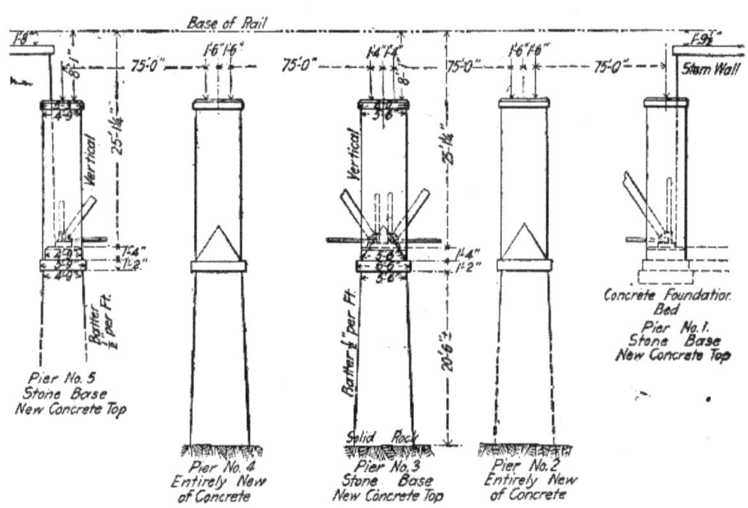

Fig. 94.—Diagram Arrangement of Piers, Calf Killer River Bridge.

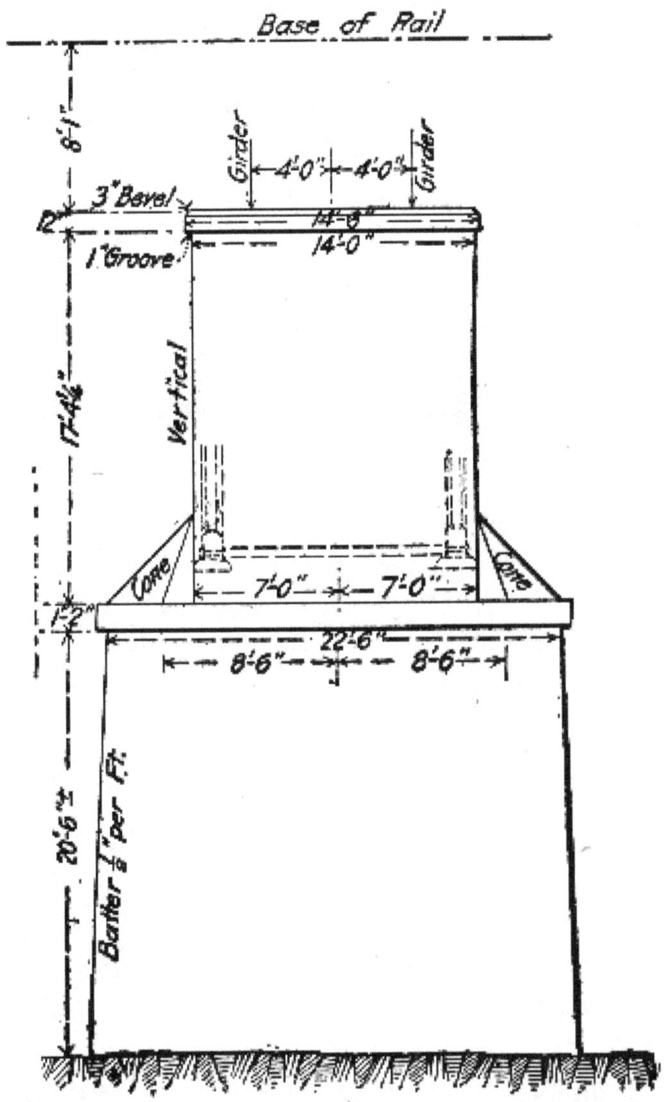

Fig. 95.—Details of Pier for Calf Killer Elver Bridge.

Slag and sand were delivered to the mixer by means of wheelbarrows. The mixer was so placed that it would dump onto a platform, and the concrete could then be shoveled into a specially designed narrow-gage car. This car ran on one rail of the main track and an extra rail outside. A turnout for clearing passing trains was provided at both ends of the bridge. The track over the bridge from the mixer had a descending grade of about 1 per cent., so that with a little start the concrete car would roll alone down to the required points on the bridge. Only in returning the empty cars to the mixer was it necessary to push them by hand, and then only for a distance of never more than 400 ft.

Over the piers on the bridge in the center of the concrete car track openings were sawed to let the concrete pass to the forms below. To get the concrete into the forms, there were used zig-zag chutes with arms about 10 ft. long, which sections were removed as the concrete in the forms was increased. These chutes were a convenience by their ends alternating from one side to the other as the arms were removed in coming up.

The cost of the concrete work was as follows:

	Rate per day.	Total days worked.	Total.	Per cu. yd. concrete.
	Unloading Material.			
Foreman	\$3.40	5	\$17.00	\$0.04
11 laborers	1.36 8/10	52	71.14	.15
Total for unloading material				\$0.19
	Building Forms, Bins, Etc.			
Foreman	\$3.40	18	\$61.20	\$0.14
9 carpenters	2.25	166	373.50	.81
New lumber, 23.7 M. ft. at \$17.80			421.86	.92
Old lumber, 6 M. ft. at \$8.33			49.98	.11
Total for building forms, bins, etc.				\$1.98
	Cofferdam Excavation (45 cu. yds.)			
Foreman	\$3.40	8	\$27.20	\$0.06
9 laborers	1.15 6/10	74½	86.12	.19
Total for cofferdam excavation				\$0.25
	Cofferdam Concrete (37 cu. yds.)			
Foreman	\$3.40	8	\$27.20	\$0.06

11 laborers	$1.36^{3}/_{10}$	79 107.68	.23
Cofferdam lumber, 2.25 M. ft. at $20.00		45.00	.09
Total for cofferdam concrete			$0.38
	Concrete Mixing and Placing.		
Foreman	$3.40	30 $102.00	$0.22
9 laborers	$1.15^{6}/_{10}$	282 325.99	.71
Cement, 452 bbls. at $1.55		701.00	1.52
Slag, 437 cu. yds. at $0.20		87.40	.19
Sand, 220 cu. yds. at $0.30		66.00	.14
Total for mixing and placing			$2.78
	Taking Down Forms and Clearing Up.		
Foreman	$3.40	13 $44.20	$0.09
11 laborers	1.17	1.43 107.31	.36
Total for taking down forms, etc.		$200.00	$0.45
Engineering and supervision			.43
Grand total, 460 cu. yds. concrete			$6.46

The wages given are the average wages. The men worked a 10-hour day. The concrete was a 1-3-6 mixture. The cofferdam work was done in connection with the construction of the fourth pier, this pier being the only one coming in the bed of the river to be built entirely new. The work on this was started in water about 6 ft. deep. The 37 cu. yds. of concrete is included in the total of 460 cu. yds. in the above tabulation. By itself the cost of the cofferdam work, not including cost of cement, sand and slag was as follows:

	Total.	Per cu. yd. Concrete.
Lumber	$ 45.00	$1.21
Labor, excavating	113.32	3.06
Labor, concrete	134.88	3.64
Total 37 cu. yds. concrete		$7.91

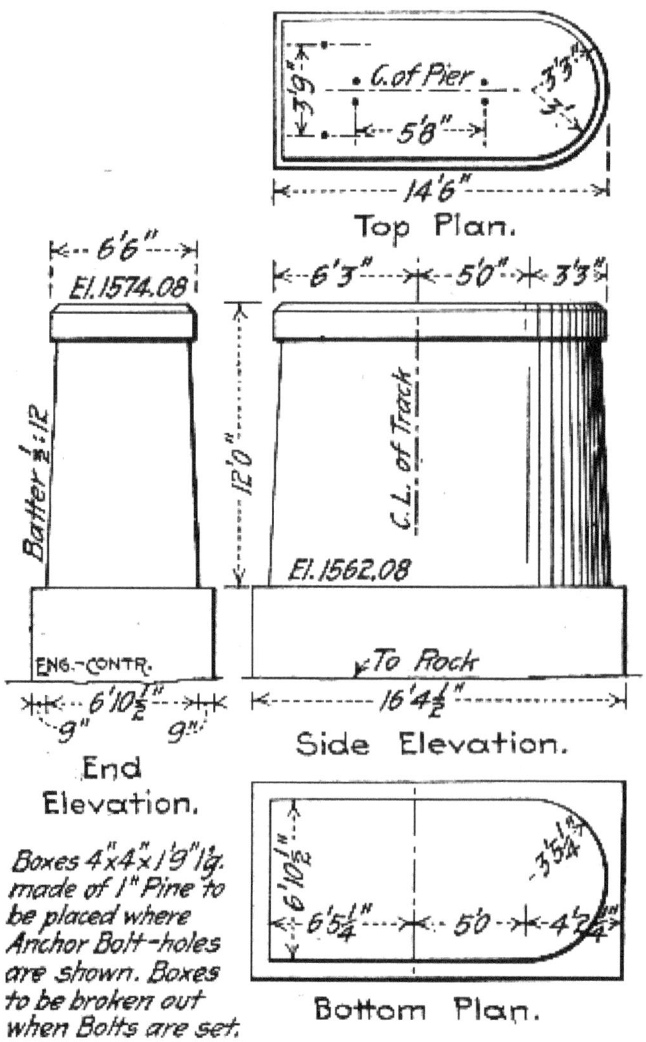

Fig. 96.—Details of Piers for K. C., M. & O. Ry. Bridge.

METHOD AND COST OF CONSTRUCTING 21 BRIDGE PIERS.— The following account of the methods and cost of constructing 21 concrete piers for a railway bridge consisting of 20 50-ft. plate girder spans has been compiled from records kept by Mr. W. W. Colpitts, Assistant Chief Engineer, Kansas City, Mexico & Orient Ry. The shape and dimensions of the piers are shown by Fig. 96 and Fig. 97 shows the construction of the forms. Sheet pile cofferdams to solid rock were used for constructing the foundations.

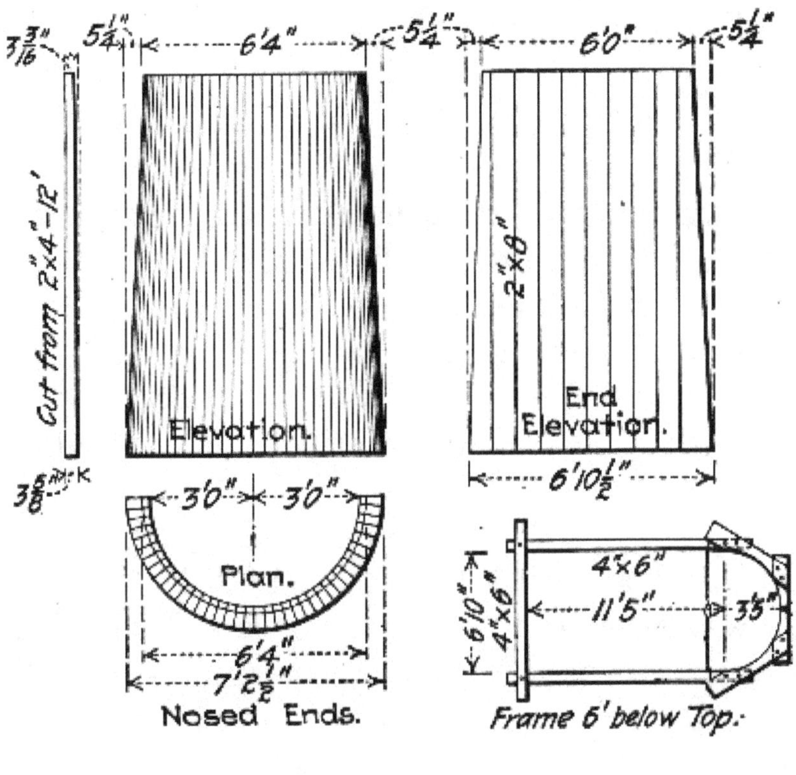

Fig. 97.—Forms for Piers for K. C., M. & O. Ry. Bridge.

The 1-3-5 concrete was mixed in a Smith mixer having a batch capacity of 9 cu. ft. The mixer was located on the slope of the embankment approach, with the main track at its rear and facing a temporary material track. This temporary track turned out from the main track about 500 ft. beyond the mixer and extended diagonally down the embankment approach on a 3 per cent. grade and across the river bottom alongside the pier sites. The portion of the track in the river bottom was supported on bents of spliced ties, jetted to the rock, and wired to the cofferdam to avoid the danger of loss in case of high water. The sand and crushed rock were delivered by cars from the main line track, immediately above the mixer, and the cement was stored in a shanty at one side of the mixer. The concrete materials and machinery were, in this manner, very conveniently located for rapid work and well above the high water line. The concrete was transported to the pier sites in improvised dump boxes, set on push cars. These dump boxes were hinged longitudinally and discharged directly into the cofferdams. The grade of the temporary track carried the push cars by gravity to the cofferdams and they were returned by teams, for which purpose a straw and brush road had been built paralleling the track. As the work progressed farther into the stream, more cars were added properly to balance the work. While the concrete in the base was still fresh, a number of steel reinforcing bars, 8 ft. in length, were set in place along each end to insure a good bond between the base and shaft.

In general, the work of putting in the bases was organized so that about the same time was required in filling a cofferdam with concrete, in excavating the sand from the next, and in driving the sheet piling for the third. These three operations were thus carried on simultaneously and, although interruptions in one part of the work or the other occurred frequently, the gangs were interchangeable and no appreciable loss was suffered, except in time, because of such delays.

In piers 19 and 20, where the rock was from 17 to 19 ft. below the surface, some difficulty was encountered due to the presence of fissures in the rock, from which it was necessary to remove the sand to fill with concrete. In such cases, the larger leaks were stopped as much as possible by driving sheet piles against the outside face of the cofferdam and into the fissures, and the smaller leaks by manure in canvas bags rammed into the openings.

Upon the completion of all the bases, the forms for several shafts were set in position and the work of filling with concrete proceeded as in the case of the bases, except that a derrick erected on a flat car and stationed at the pier was utilized to raise the dump boxes in depositing the concrete in the forms. As soon as the concrete in one shaft had set sufficiently to permit of it, the forms were removed and placed on the pier ahead. Four sets of forms were used for the shafts.

The following are the average prices paid for materials and labor:

Materials.—Lumber for forms, etc., \$16.50 per M. ft., B. M.; cement, Kansas Portland, \$1.50 per bbl.; broken limestone, 45c per cu. yd.; sand, Arkansas River, 15c per ton.

Labor.—General foreman, \$110 per month; assistant foreman, \$75 per month; timekeeper, \$60 per month; riveters, 35c per hour; blacksmith, 30c per hour; blacksmith assistant, 20c per hour; carpenters, 22½c and 25c per hour; enginemen, 25c per hour; firemen, 20c per hour; night watchman, 20c per hour; laborers, 17½c and 20c per hour; team (including driver), 40c per hour. The prices quoted for lumber, cement, limestone and sand are prices f. o. b., Louisiana, Iola, Kan., El Dorado, Kan., and Wichita, Kan.

The total and unit cost of constructing the concrete piers and abutments and of erecting the steel superstructure are given in the following tabulation. Altogether there was about 2,300 cu. yds. of concrete in the substructure, most of which, as stated above, was a 1-3-5 mixture.

Machinery and Supplies.	
Concrete mixer, 20% of cost	\$ 152.10
Supplies, freight, hauling, setting up	505.04
Total	\$ 657.14
Centrifugal sand pump, 20% of cost	\$ 27.00
Supplies, freight, hauling, setting up	277.50
Rent of traction engine to operate	83.25
Total	\$ 387.75
Water pump and pipe, 20% of cost	\$ 29.00
Supplies, freight, hauling, setting up	177.32
Total	\$ 206.32
Pile driver engine, 20% of cost	\$ 100.00
Supplies, freight, hauling, setting up	243.65
Total	\$ 343.65

Grand total	$1,594.86
Cofferdams.	
Materials, lumber and nails	$1,285.26
Freight and train haul	306.33
Labor making piles	696.82
Labor driving piles	1,384.05
Total	$3,672.46

The sheet piling took 63,500 ft. B. M. of lumber; the cost per 1,000 ft. B. M. for the sheet piling was then:

Materials, lumber and nails	$ 20.08
Freight and haulage	4.82
Labor making piles	10.97
Labor driving piles	21.80
Total	$ 57.67
Forms, Platforms and Runways.	
Lumber, hardware, etc.	$224.59
Freight and train haul.	40.20
Labor making, removing and placing.	556.51
Total	$821.30
Concrete Materials.	
Cement, freight, unloading and storing.	$4,617.48
Sand, freight, unloading, etc.	1,336.05
Broken stone, freight, unloading, etc.	2,026.92
Total	$7,980.45

This gives us for 2,300 cu. yds. of concrete a cost of $3.47 per cu. yd. for materials, including freight, storage, and unloading charges of all kinds. A line on the proportion of the cost contributed by these latter items may be got by taking the prices of the materials f. o. b. at the places of production and assuming the proportions for a 1-3-5 concrete. According to tables in Chapter II, a 1-3-5 broken stone concrete requires per cubic yard 1.13 bbls. cement, 0.48 cu. yd. sand and 0.80 cu. yd. broken stone. We have then:

1.13 bbls. cement, at $1.50	$1.69
0.48 cu. yd. sand, at 20c	.10
0.80 cu. yd. stone, at 45c	.36
Total	$2.15

This leaves a charge of $1.32 per cubic yard of concrete for freight and handling materials. The cost of mixing concrete and placing it in the forms was $3,490.87, or $1.52 per cu. yd. We have then:

Cost of concrete materials per cu. yd.	$3.47
Cost of mixing and placing concrete.	1.52
Total.	$4.99

The miscellaneous expenses of the work comprised:

Watchman, tools, telephone, etc.	$ 722.48
Shanties, furnishings, supplies, etc.	829.04
Total.	$1,551.52

To this has to be added $1,134.28, the cost of excavating the cofferdams. The total and unit costs of the different items of the concrete substructure work can now be summarized as follows:

Item.	Total.	Per cu. yd.
Machinery and supplies	1,594.86	$ .69
Cofferdams	3,672.49	1.60

Forms, etc	821.30	.36
Concrete materials	7,980.45	3.47
Mixing and placing concrete	3,490.87	1.53
Excavating cofferdams	1,134.28	.49
Miscellaneous	1,551.52	.67
Total	$20,245.74	$8.81

COST OF PERMANENT WAY STRUCTURES KANSAS CITY OUTER BELT & ELECTRIC RY.—The following cost of concrete work including retaining walls, abutments and box culverts, for the permanent way of the Kansas City Outer Belt & Electric Ry., is given by Mr. W. W. Colpitts. These figures are of particular interest, for the variation in prices of materials during the two-year period while work was in progress and as giving the average cost of the work on the whole line as well as for individual structures. The culverts were all box culverts with wing walls and the abutments were for girder bridges. Walls and abutments were of L section with triangular or trapezoidal counterforts at the back between base slab and coping. The form work was thus rather complex.

All work was reinforced concrete, and was done by contract under the following conditions: The work of preparing foundations, including excavation, pile driving, diversions of streams, etc., was done by the railroad company, which also bore one-half the cost of keeping foundations dry while forms were being built and concrete placed. The railroad company also furnished the reinforcing bars at the site of each opening. The concrete work was let at $9 per cu. yd., which figure covered all the labor and materials necessary to complete the work, other than the exceptions mentioned. The concrete proportions were 1-3-5. The cement used was Iola Portland and Atlas Portland. The sand was obtained from the bed of the Kansas River in Kansas City. The rock used was crushed limestone, passing a 2-in. ring and freed from dust by screening. Corrugated reinforcing bars, having an elastic limit of from 50,000 to 60,000 lbs. per sq. in., manufactured by the Expanded Metal & Corrugated Bar Co. of St. Louis, Mo., were used exclusively. The concrete in the smaller structures was mixed by hand, in the larger by a No. 1 Smith mixer. In the first structures built 2-in. form lumber was used, with 2 by 6-in. studs placed 3 ft. on centers. This was abandoned later for 1-in. lumber with 2 by 6-in. studs, 12 ins. on centers, and was found to be more satisfactory in producing a better face. The structures were built in the period from April, 1905, to May, 1907.

The cost of materials and the wages paid labor were as follows:

Cement—	
Per barrel at structure, April, 1905	$1.25
Per barrel at structure, April, 1907	1.92
Average cost per barrel at mill	1.42
Freight per barrel	0.21
Hauling 1½ miles and storage	0.12
Average cost at structure	1.75
Average cost per cu. yd. concrete (1.1 bbls.)	1.93
Sand—	
Per cu. yd. at structure, April, 1905	$0.625
Per cu. yd. at structure, April, 1907	0.75
Average cost per cu. yd., river bank	0.30
Freight per cu. yd	0.22
Hauling 1½ miles	0.20
Average cost at structure	0.72
Average cost per cu. yd. concrete (½ cu. yd.)	0.36
Stone—	
Per cu. yd. at structure, April, 1905	$1.10
Per cu. yd. at structure, April, 1907	1.75
Average cost per cu. yd. at crusher	0.65
Hauling 4 miles	0.84
Average cost at structure	1.49
Average cost per cu. yd. concrete (0.9 cu. yd.)	1.34
Lumber—	
Per M. ft. at structure, April, 1905	$15.00

Per M. ft. at structure, April, 1907	22.50	
Average cost per M. at structure	19.00	
Average cost per cu. yd. concrete	0.49	
Labor—	Max.	Min.
Common labor, cts. per hour	20	17
Carpenters, cts. per hour	40	30

With these prices and wages the average cost of concrete work for the whole line was:

Item.	Per cu. yd.
Form building and removing	$1.98
Mixing and placing concrete	0.74
Placing reinforcement	0.10
Wire, nails, water, etc.	0.20
1.1 bbls. cement at $1.75	1.93
½ cu. yd. sand at $0.72	0.36
0.9 cu. yd. stone at $1.49	1.34
Lumber for forms	0.49
Total	$7.14

The following are the costs of specific structures built at different times:

Example I.—Indian Creek Culvert. 14×15 ft., 250 long, completed November, 1905:

	Per cu. yd.
Cement	$1.37
Sand	.34
Stone	1.10
Labor	2.48
Lumber	.76
Miscellaneous	.18

Total	$6.23

Example II.—Third Street Abutments and Retaining Wall. Completed November, 1906:

	Per cu. yd.
Cement	$1.78
Sand	.35
Stone	1.35
Lumber	.74
Labor	2.75
Miscellaneous	.16
Total	$7.13

Example III.—Abutments, Overhead Crossing with Union Pacific and Rock Island. Completed May, 1907:

	Per cu. yd.
Cement	$1.92
Sand	.32
Stone	1.74
Lumber	.98
Labor	2.96
Miscellaneous	.16
Total	$8.08

COST OF PLATE GIRDER BRIDGE ABUTMENTS.—The following record of the construction of 20 abutments for 10 four-track plate girder bridges over streets in Chicago, Ill., are given by Mr. W. A. Rogers. The work was done between May 1 and Oct. 1, 1898, in which time 8,400 cu. yds. of concrete were placed, all the work being done by company labor.

The forms were made of 2-in. plank and 6×6-in. posts bolted together at the top and bottom with ¾-in. rods. The lumber was used over and over again. When the dressed plank became too poor for the face it was used for the back. The concrete was 1 Portland cement, 3 gravel and 4 to 4½ limestone (crusher run up to 3-in. size). A mortar face 1½ ins. thick was built up with the rest of the concrete. The concrete was made quite wet, and each man ramming averaged 18 cu. yds. a day rammed. The concrete was mixed by a machine of the Ransome type, operated by a 12-HP. portable gasoline engine. The load was very light for the engine, and 8 HP. would have been sufficient. The engine made 235 revolutions per minute, and the pulley wheels were proportioned so that the mixer made 12 revs, per minute. One gallon of gasoline was used per hour, and the mixing was carried on day and night so as not to give the concrete time to set. The time required for each batch was 2 to 3 mins., and about ½ cu. yd. of concrete was delivered per batch. The average output was 70 cu. yds. per 10-hr, shift, with a crew of 28 men; but as high as 96 cu. yds. were mixed in 10 hrs. The concrete was far superior to hand mixed concrete. The water for the concrete was measured in an upright tank and discharged by a pipe into the mixer. The sand and stone were delivered to the mixer in wheelbarrows, and the concrete was taken away in wheelbarrows. No derricks were used at all. Each wheelbarrow of concrete was raised by a rope passing over a pulley at the top of a gallows frame, one horse and a driver serving for this raising. A small gasoline hoisting engine would have been more satisfactory than the horse which was worked to its full capacity. After the barrows were raised (12 ft.), they were wheeled to the abutment forms and dumped. The empty wheelbarrows were lowered by hand, by means of a rope passing over a sheave and provided with a counterweight to check the descent of the barrow. The cost of the concrete (built by company labor) was as follows:

	Per cu. yd.
Cement, gravel and stone delivered	$3.28
Material in forms (used many time)	.11
Carpenters building and taking down forms	.34
Labor	1.18
Total per cu. yd	$4.91

The labor cost includes moving plant from one bridge to the next, building runways, gasoline for engine, oil for lights at night and unloading materials, as well as mixing, transporting and placing concrete. Wages were $1.75 per 10-hour day for laborers and $2.50 for carpenters.

COST OF ABUTMENTS AND PIERS, LONESOME VALLEY VIADUCT.—Mr. Gustave R. Tuska gives the following on the concrete substructure of the Lonesome Valley Viaduct, near Knoxville, Tenn. There were two U-shaped abutments and 36 concrete piers made of a light limestone that deteriorates rapidly when used for masonry. Derricks were not needed as would have been the case with masonry piers, and colored labor at $1 for 11 hrs. could be used. The piers were made 4 ft. square on top, from 5 to 16 ft. high, and with a batter of 1 in. to the foot. The abutments average 26 ft. high, 26 ft. long on the face, with wing walls 27 ft. long; the wall at the bridge seat is 5 ft. thick, and the wing walls are 3½ ft. wide on top. Batters are 1 in. to the foot.

The forms were made of 2-in. tongued and grooved plank, braced by posts of 2×10-in. plank placed 3 ft. c. to c. for the abutments, and at each corner for the piers. At the corners one side was dapped into the other, so as to prevent leakage of cement. The posts were braced by batter posts from the earth. For the piers a square frame was dropped over the forms and spiked to the posts. The abutment forms were built up as the concreting progressed. The north abutment forms were made in sections 6 ft. high, held by ¾-in. bolts buried in the concrete. The lower sections were removed and used again on the upper part of the work, thus saving plank. The inside of forms was painted with a thin coat of crude black oil. The same form was used for several piers.

The concrete was 1-2-5, the barrel being the unit of measure, making about ¾ cu. yd. of concrete per batch. The mortar was mixed with hoes, but shovels were used to mix in the stone. By passing the blade of a shovel between the form and the concrete, the stone was forced back and a smooth mortar face was secured. Rammers weighing 30 to 40 lbs. were used for tamping. Two days after the completion of a pier the forms were removed. The concrete was protected from the sun by twigs, and was watered twice a day for a week. It was found by actual measurement that 1 cu. yd. Of concrete (1-2-5), the ingredients being measured in barrels, consisted of 1¼ bbls. of Atlas cement, 10 cu. ft. of sand, and 26½ cu. ft. of stone. The total amount of concrete was 926 cu. yds. of which two-thirds was in the two abutments. The work was done (in 1894) by contract, for $7 per cu. yd., cement costing $2.80 per bbl., sand 30 cts. per cu. yd., and wages $1 a day. A slight profit was made at this price. A gang of 15 men and a foreman would mix and lay about 40 cu. yds. in 11 hrs. when not delayed by lack of materials. The cost of making the concrete, with wages at $1 a day, was:

	Cts. per cu. yd.
1 man filling sand barrels and handling water	2.7
2 men filling rock barrels	5.4
4 men mixing sand and cement	10.6
4 men mixing stone and mortar	10.6
2 men wheeling concrete	5.3
1 man spreading concrete	2.7
1 man tamping	2.7
1 foreman	5.0
Total labor	45.0

COST OF HAND MIXING AND WHEELBARROW WORK FOR FOUR BRIDGE PIERS.—The following figures of the cost of hand-mixed concrete for bridge piers and abutments are given by Mr. Fred R. Charles of Richmond, Ind. The figures cover three jobs. All concrete was mixed by hand and with one exception noted below was moved to place in wheelbarrows. The concrete was a 1-2½-5½ mixture. In this connection it is well to note that in one or two of the jobs where the proportion of the aggregate seems too small for the yardage of concrete the difference is accounted for by the fact that large stones were placed in the foundations, these stone being on the ground and costing nothing but the labor to throw them in.

Job I.—The first job consisted of the construction of one abutment and six piers for a bridge over the Miami River at Fernald, O. The stone was procured on the site and crushed by a portable crusher run by a traction engine. The rough stone cost 10 cts. a cubic yard, and this, with the cost of handling, fuel and hire of engine and crusher, made the cost of crushed stone about $1 per cu. yd. Sand was obtained close to the work, but the cement had to be teamed 10 miles. Labor was paid $1.75 per day. The cost of materials and labor per cubic yard of concrete in place was as follows:

Item.	Per cu. yd.
1.16 bbls. cement at $2.10	$1.58
Sand	0.35
Stone	0.75
Lumber	0.64
Tools, hardware, etc.	0.20
Labor (including 15 cts. per cu. yd. for pumping)	2.78
Total materials and labor	$6.30

Job II.—The second job was the construction of two abutments containing 434 cu. yds. of concrete for a viaduct at Ernst Street, Cincinnati, O. The abutments were constructed at the street and the excavation was clay and shale. Labor received $1.75 per day. The cost of materials and labor per cubic yard of concrete in place was as follows:

Materials—	Per cu. yd.
376 bbls. cement at $1.70	$1.48
224 cu. yds. sand at $1.20	0.64
255 cu. yds. stone at $1.55	1.00
Lumber	0.40
Tools, hardware, etc.	0.06
Total materials	$3.58
Labor—	
Clearing and excavating	$1.12
Mixing and placing concrete	1.13
Building forms, etc.	0.25
Total labor	$2.50

Total labor and materials	$6.08

Job III.—This job consisted in placing 570 cu. yds. of concrete in the pedestals for a viaduct at Quebec Avenue, Cincinnati, O. The pedestals were 5 ft. square on top and from 8 to 20 ft. high. The location of the work was very inconvenient for the delivery of materials, all materials having to be teamed or wheeled. Labor was paid $1.75 per day. The cost of labor and materials per cubic yard of concrete in place was as follows:

Item.	Per cu. yd.
500 bbls. cement at $1.60	$1.40
239 cu. yds. sand at $1.25	0.53
560 cu. yds. stone at $1.88	1.84
Lumber	0.38
Tools, hardware, etc.	0.05
Labor	2.96
Total labor and materials	$7.16

Job IV.—This job consisted in placing 2,111 cu. yds. of concrete in a railway viaduct at Cincinnati, O. For one pier 56 ft. high the concrete was raised to place by a derrick; for the remainder of the work it was wheeled or teamed to place. Labor was paid $1.75 per day. The cost of labor and materials per cubic yard of concrete in place was as follows:

Item.	Per cu. yd.
1,908 bbls. cement at $1.60	$1.44
1,105 cu. yds. sand at $1.95	0.50
1,468 cu. yds. stone at $1.48	1.03
Lumber	0.54
Tools, hardware, etc.	0.25
Water	0.03
Labor	3.44
Total labor and materials	$7.23

13

RETAINING WALLS CONSTRUCTION

Concrete retaining walls may for construction purposes be divided into two classes: Plain concrete walls of gravity section and reinforced concrete walls consisting of a thin slab taking the thrust of the earth as a cantilever anchored to a base slab or as a flat beam between counterforts. The reinforced wall requires much less concrete for a given height than does the plain, gravity wall, but the concrete is more expensive owing to the reinforcement and to the more complex form of construction, and, in some measure, to the greater cost of placing the mixture in narrow forms and around reinforcement. It is common, too, to require a richer concrete for the reinforced than for the plain wall.

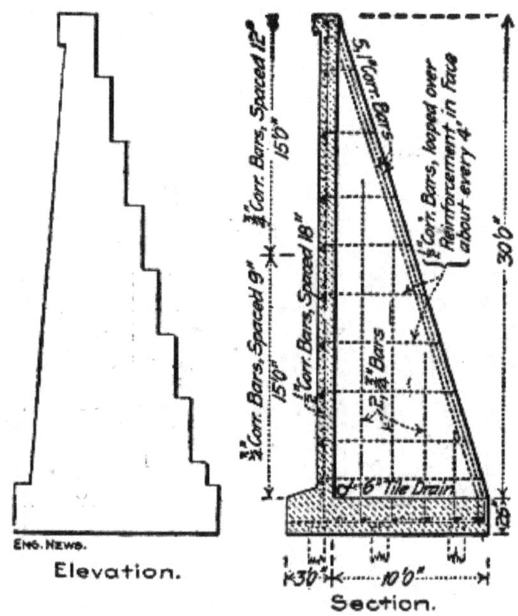

Fig. 98.—Comparison of Plain and Reinforced Sections for Retaining Walls (C. E. Graff).

COMPARATIVE ECONOMY OF PLAIN AND REINFORCED CONCRETE WALLS.—Prior to the construction of some 2,000 ft. of retaining wall ranging in height from 2 ft. to 38 ft., at Seattle, Wash., calculation was made by the engineers of the Great Northern Ry. to determine the comparative economy of plain concrete and reinforced concrete sections. The sections assumed were those shown by Fig. 98, and comparisons were made at heights of 10, 20, 30 and 40 ft., with the following results:

Height in feet.	Plain. Cu. yds. per ft.	Reinforced. Cu. yds. per ft.	Per cent. Saving.
10	1.63	1.29	20.4
20	4.08	2.59	36.4
30	8.40	4.73	43.3
40	14.70	8.07	45.0

The saving in concrete increased as the height of the wall increased; for a 40-ft. wall reinforced concrete at nearly double the cost per cubic yard in place would be as cheap as plain concrete.

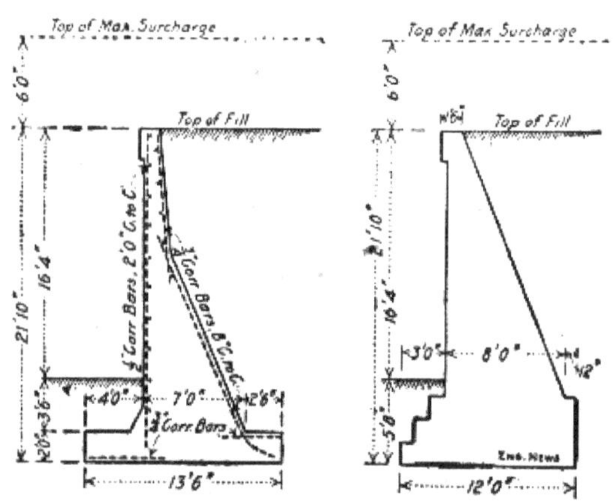

Fig. 99.—Comparison of Plain and Reinforced Sections for Retaining Wall (F. F. Sinks).

Taking substantially the section of reinforced wall being used on the Chicago track elevation work of the Chicago, Burlington & Quincy R. R., and comparing it with a plain wall as shown by Fig. 99, Mr. F. F. Sinks obtained the following results:

Plain Wall, Cost per Lineal Foot—

4.8 cu. yds. concrete at $4	$19.20
115 ft. B. M. of forms at $31	3.56
Total 4.8 cu. yds. at $4.74	$22.76

Reinforced Wall, Cost per Lineal Foot—

3.46 cu. yds. concrete at $4.10	$14.18
115 ft. B. M. of forms at $31	3.56

109 lbs. reinforcing steel at 3¼ cts.	3.54
1.34 cu. yds. extra fill at 20 cts.	0.27
0.32 cu. yd. extra excavation at 20 cts.	0.06
Total, 3.46 cu. yds. concrete at $6.25	$21.61

The saving in this case was $1.15 per lineal foot of wall with the unit cost of reinforced concrete in place 24 per cent. greater than the unit cost of plain concrete. It will be noted that there is some 28 per cent. less concrete per lineal foot of wall in the reinforced section and also that this section is so designed that the form work is about as simple for one section as for the other. Another point to be noticed is that there is no saving in excavation by using a reinforced section instead of a gravity section, in fact the excavation runs slightly more for the reinforced section.

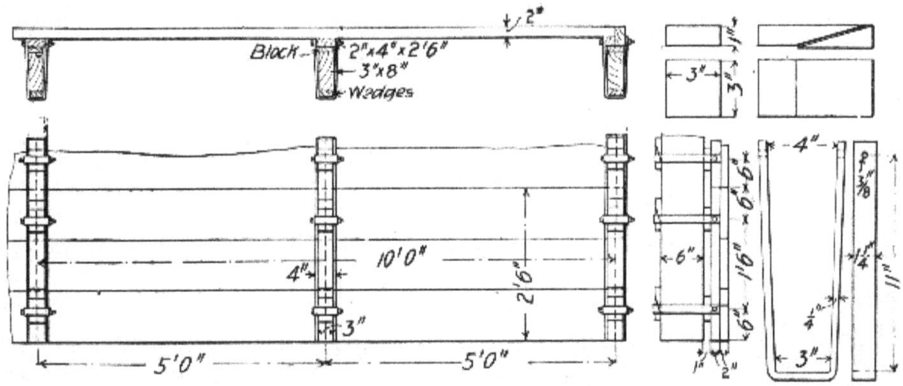

Fig. 100.—Forms for Retaining Wall Work, N. Y. C. & H. R. R. R.

FORM CONSTRUCTION.—Retaining wall work often affords an opportunity for constructing the forms in panels and this opportunity should be taken advantage of when possible. Several of the walls described later give examples of form work that may be studied with profit in this respect.

Figure 100 shows a panel form construction employed on the New York Central & Hudson River R. R. The 3×8-in. studs are erected, care being taken to get them in proper line and to true batter and also to brace them rigidly by diagonal props. Generally the studding is erected for a section of wall 50 ft. long at one time. The lagging, made in panels 2½ ft. wide and 10 ft. long, by nailing 2-in. plank to 2×4-in. cleats, is attached to the studding a panel at a time and beginning at the bottom, by means of the straps, wedges and blocks shown. Five bottom panels making a form 2½ ft. high and 50 ft. long are placed first. When the concrete has been brought up nearly to the top of these panels, a second row of panels is placed on top of the first. When it is judged that the concrete is hard enough the lowermost panels are loosened and made free by removing the wedges, blocks and straps and the panels are drawn out endwise from behind the studding and used over again for one of the upper courses. The small size of the panels makes it practicable to lay bare the concrete while it is yet soft enough to work with a float or to finish by scrubbing as described in Chapter VIII. In cases where this object is not sought, panels of much larger size may be used. Working with panels 2¼×12½ ft. of 2-in. plank it was found that each panel could be used 16 times before becoming unfit for further use, but as, owing to the nicety of molded surface demanded, panels were discarded when showing comparatively small blemishes, this record cannot be taken as a true indication of the life of such forms. These panel forms are used by the railway named for long abutments and piers as well as for retaining walls.

A different type of sectional form construction is illustrated by Figs. 101 and 102. It has been extensively used for retaining wall work by the Chicago, Burlington & Quincy R. R. The studding and waling are framed in units as shown. The lagging is framed in panels for the rear of the wall, for the face of the coping, and for the inclined toe of the wall, and is ordinary sheathing boards for the main face of the wall. The make-up of the several panels is shown by the drawings. The reason for using ordinary sheathing instead of panels for the face of the wall is stated by Mr. L. J. Hotchkiss, Assistant Bridge Engineer, to be that "the sections become battered and warped with use, do not fit closely together, and leave the wall rough when they are removed." The manner of bracing the form and of anchoring it down against the up-thrust of the wet concrete is shown by Fig. 102.

Two other examples of sectional form construction are given in the succeeding descriptions of work for the Grand Central Station terminal in New York City and for the Chicago Drainage Canal. In the former work it is notable that panels 51×20 ft. were used, being handled by locomotive crane. The panels used on the drainage canal work and in the forms previously described are of sizes that can be taken down and erected by hand, and the means of handling them should always be given consideration in deciding on the sizes to be adopted for form panels not only in wall construction but in any other class of work where sectional forms may be used. Wet spruce or yellow pine will weigh 4½ lbs. per ft. B. M., so that a panel 10×2½ ft.

made of 2-in. plank and three 2×4-in. battens will weigh some 225 lbs. In form work where the panels are removed and re-erected in succession facility in handling is an important matter. When one figures that he may handle both the concrete and the form panels with it a cableway or a locomotive crane becomes a tool well worth considering in heavy wall work.

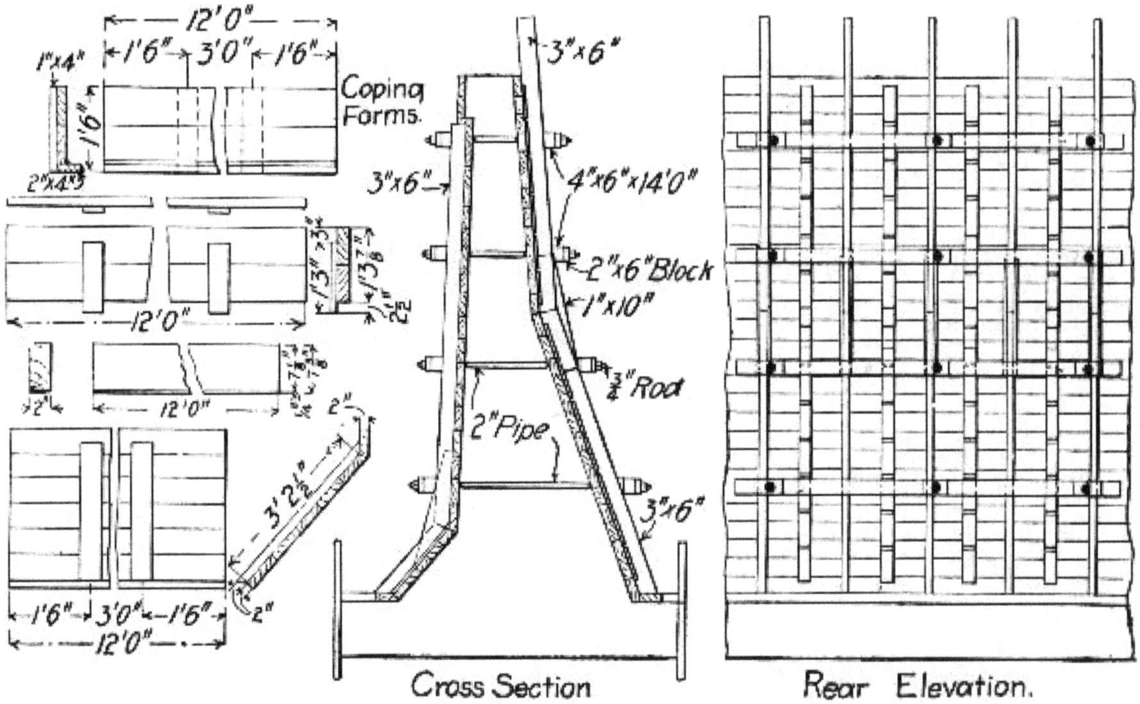

Fig. 101.—Forms for Retaining Wall Work, C., B. & Q. R. R.

Three details in retaining wall form work that are often sources of annoyance out of proportion to their magnitude are alignment, coping construction and wall ties. Small variations from line in the face of the wall are seldom noticeable, but a wavy coping shows at a glance. For this reason it is often wise to build the coping after the main body of the wall has been stripped, or if both are built together to provide in the forms some independent means of lining up the coping molds. In the form shown by Fig. 101 the latter is done by bracing the coping panel so as to permit it to be set and lined up independently of the main form. A separate form for molding the coping after the main body of the wall is completed may be constructed as shown by Fig. 103. Bolts at B and C permit the yokes to be collapsed and the form to be shifted ahead as the work advances. This mold provides for beveling the top edges of the coping and also the edge of the overhang, and the beveling or rounding of these edges should never be omitted where a neat appearance is desired. It is not essential, however, that this finishing be done in the molds. By stripping the concrete while it is still pliable the edges can be worked down by the ordinary cement sidewalk edger.

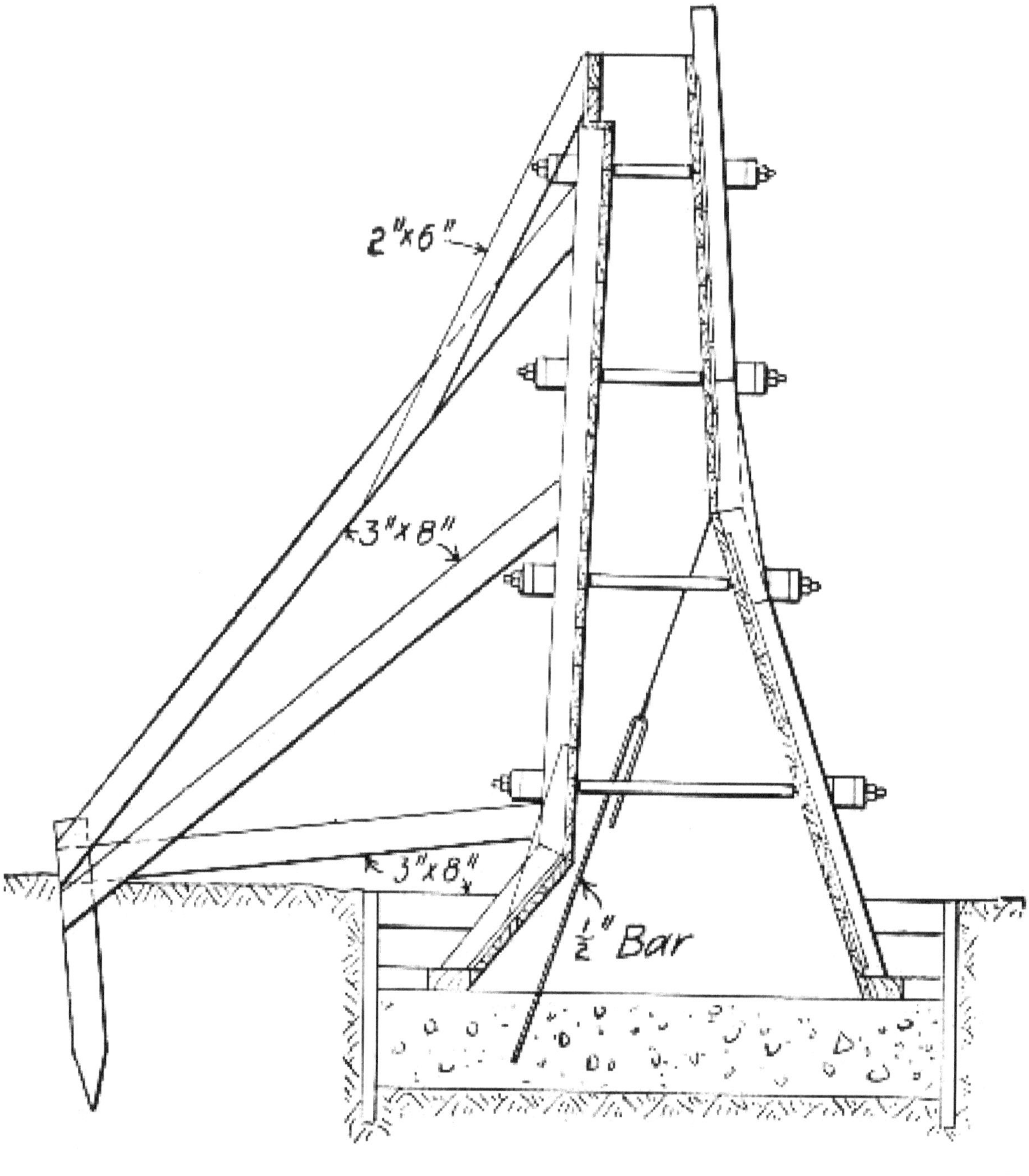

Fig. 102.—Sketch Showing Method of Bracing Form Shown by Fig. 101.

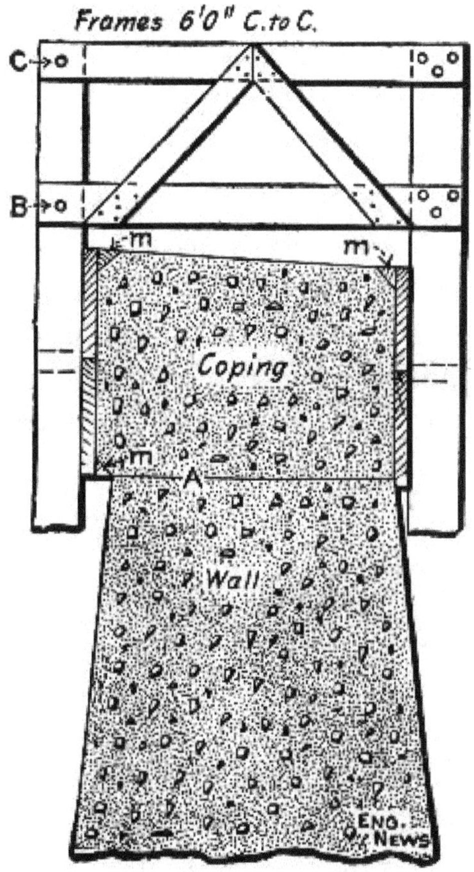

Fig. 103.—Sectional Form for Constructing Coping.

Wall ties are commonly used to hold the face and back forms to proper spacing, but occasionally they are not permitted. In the latter case the bracing must be arranged to hold the forms from tipping inward as well as from being thrust outward. A good arrangement is that shown by Fig. 102. In fastening the forms with ties the choice is usually between long bolts which are removed when the molds are taken down and wire ties which are left embedded in the concrete. The selection to be made depends upon the character of the work. When sectional forms are used like the one shown by Fig. 101, for long stretches of wall of nearly uniform cross-section bolts are generally more economical and always more secure. If the bolts are sleeved with scrap gas pipe having the ends corked with waste the bolts can be removed ordinarily without difficulty. To make the pipe sleeve serve also as a spacer the end next the face may be capped with a wooden washer which is removed and the hole plastered when the forms are taken down. With bolt ties the forms can be filled to a depth of 15 to 20 ft with sloppy concrete. This is hardly safe with wire ties unless more wire and better tieing are employed than is usual. It takes four strands of No. 10 to give the same working stress as a ½-in. threaded rod and the tieing in of four strands of wire so that they will be without slack and give is a task requiring some skill. Bolts are much more easily placed and made tight. In the matter of cost of metal left in the wall, the question is between the cost of scrap gas pipe and of wire; the pound price of the wire is greater but fewer pounds are used and the metal is in more convenient shape to cut to length and to handle. This convenience in shaping the tie to the work gives the advantage to wire ties for isolated jobs or jobs which involve a continual change in the length and spacing of the ties. In general the contractor will find bolts preferable where sectional forms are used and wire ties preferable when using continuous forms.

One objection urged against the use of wire ties is that the metal is exposed at the face of the work when they are clipped off unless the concrete is chipped and the cavity plastered. To obviate this objection various forms of removable "heads" have been devised. Two such devices are shown by Figs. 104 and 105. In both the bolt is unscrewed, leaving the "heads" embedded. The head shown by Fig. 104 has the advantage that it can be made by any blacksmith, while the head shown by Fig. 105 is a special casting.

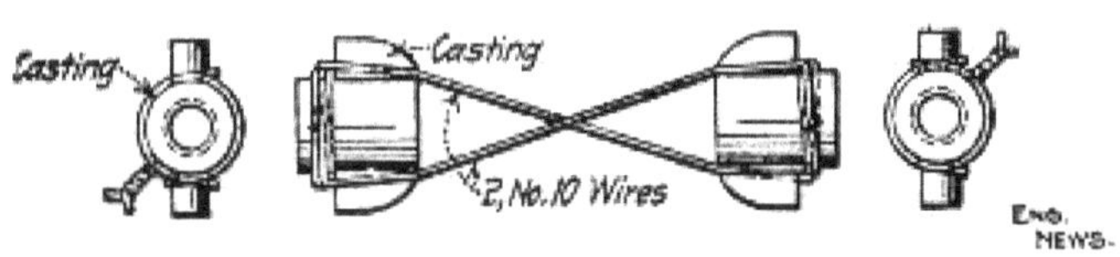

Fig. 104.—Tie for Wall Forms.

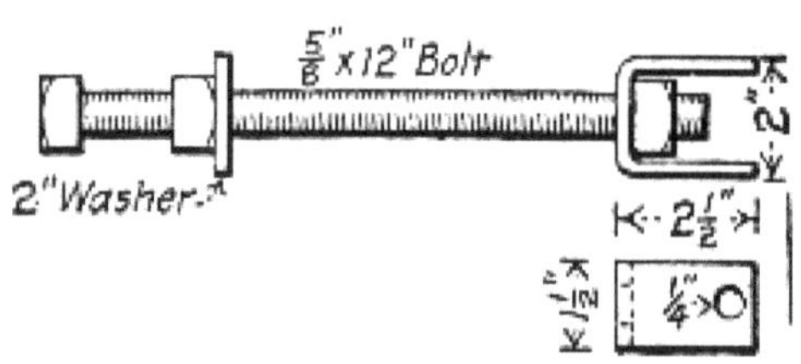

Fig. 105.—Tie for Wall Forms.

MIXING AND PLACING CONCRETE.—Where a long stretch of wall is to be built the system of mixing and handling the concrete must be capable of being shifted along the work. For isolated walls of short length this problem is a simpler one. Where the mixer can be installed on the bank above, wheeling to chutes reaching down to the work is the best solution.

As shown in Chapter IV concrete can be successfully and economically chuted to place to a greater extent than most contractors realize. Where the mixer has to be installed at the foot of the wall wheelbarrow inclines, derricks, gallows frames, etc., suggest themselves as means of handling the concrete. It is not this class of work, however, but the long stretches of heavy section walls such as occur in depressed or elevated railway work in cities that call for thought in the arrangement and selection of mixing and handling plant.

In building the many miles of retaining wall in the work of doing away with grade crossings in Chicago, Ill., trains made up of a mixer car and several material cars have been used. The mixer is mounted on a flat car set at the head of the train and is covered by a decking carrying two charging hoppers set above the mixer. The material cars are arranged behind, the sand and stone or gravel being in gondola cars.

Portable brackets hooked to the sides of the gondola cars carry runways for wheelbarrows. Sand and stone or gravel are wheeled to the charging hoppers, the work being continuous since one hopper is being filled while the other is being discharged into the mixer. The mixer discharges either into a chute, wheelbarrows or buckets. The foregoing is the general arrangement; it is modified in special instances, as is mentioned further on. The chief objection to the method is the difficulty of loading the wheelbarrows standing on runways level with the tops of the gondola sides. The lift from the bottom of the car is excessive, and as pointed out previously, shoveling stone or gravel by digging into it from the top is a difficult task.

The delivery of the concrete into the forms was accomplished by chute where possible, otherwise by wheelbarrows or cranes, and in one case by belt conveyor. In the last instance the mixer car was equipped with a Drake continuous mixer and was set in front. Behind it came three or four gondola cars of sand and stone, and at the rear end a box car of cement. All material was wheeled on side runways to two charging hoppers over the mixer. The mixer discharged onto a belt conveyor carried by a 25-ft. boom guyed to an

A-frame on the car and pivoted at the car end to swing 180° by means of a tag line. The outer end of the conveyor was swung over the forms. A ¾-in. wire rope wrapped eight times around two drums on the mixer car and passing through slots in the floor to anchors placed one 500 ft. in front and one 500 ft. to the rear enabled the train to be moved back and forth along the work. This scheme of self-propulsion saved the hire of a locomotive. In another case the mixer was discharged into buckets which were handled by a crane traveling back and forth along a track laid on two flat cars.

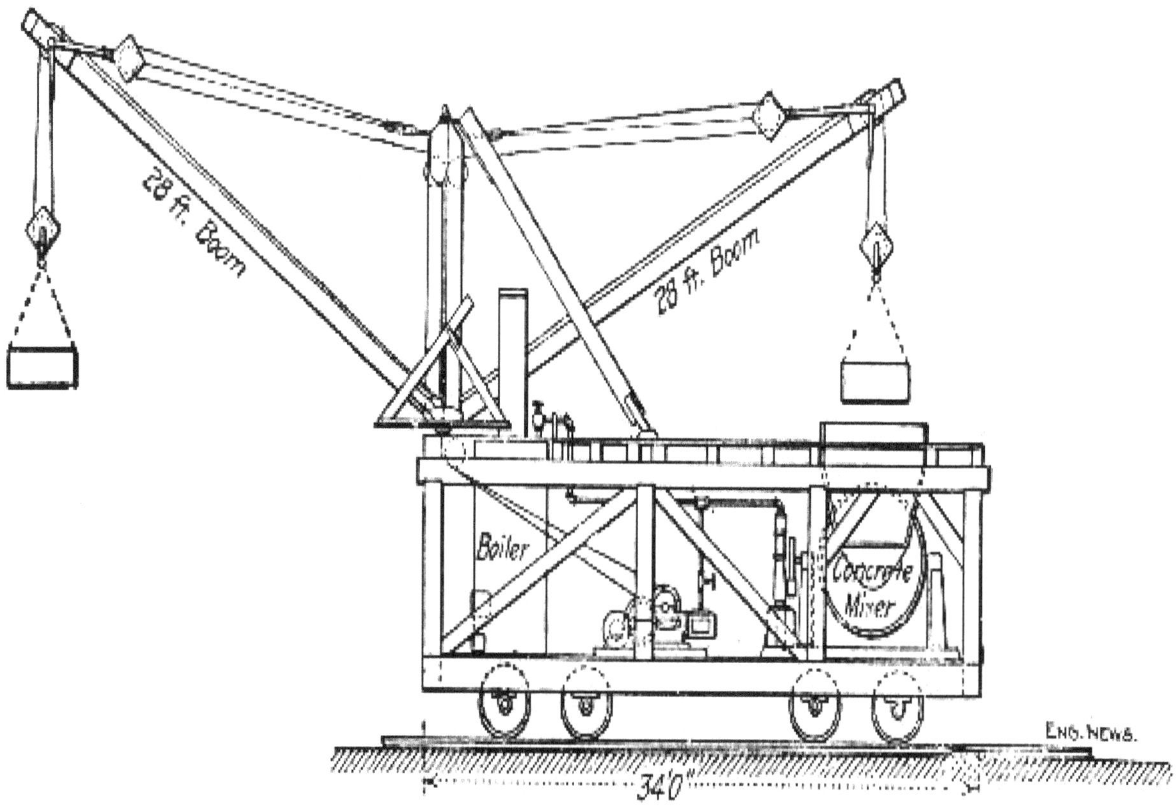

Fig. 106.—Side Elevation of Traveling Mixer Plant, Galveston Sea Wall.

Another type of movable mixer plant used in constructing a sea-wall some 3½ miles long at Galveston, Tex., is shown by Figs. 106 and 107. Two of these machines mixed and placed some 127,000 cu. yds. of concrete, in 1 cu. yd. batches. Two 12-HP. engines operated the derricks and one 16-HP. engine operated the Smith mixer; all engines took steam from a 50-HP. boiler. The rated capacity of each machine was 300 to 350 cu. yds. per day. The method of operation is clearly indicated by the drawings.

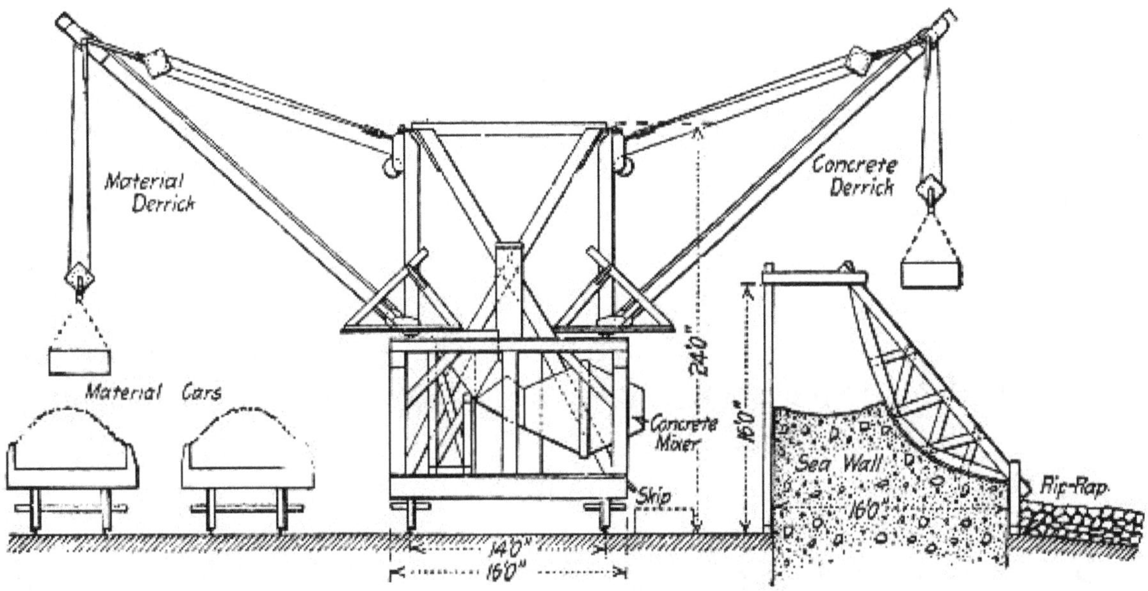

Fig. 107.—End Elevation of Traveling-Mixer, Galveston Sea Wall.

Placing the concrete in the forms is generally required to be done in layers; with wet mixtures this means little more than distributing the concrete somewhat evenly along the wall and slicing and puddling it to get rid of air and prevent segregation. Where mortar facing is required the face form described in Chapter VIII may be used. A reasonably good surface can be secured without mortar facing by spading the face. With dry concrete, placing and ramming in layers, calls for such care as is necessary in dry concrete work everywhere. Where new concrete has to be placed on concrete placed the day before, good bond may be secured and the chance of efflorescence be reduced by the methods described in Chapter VIII.

WALLS IN TRENCH.—In canal excavation, in subway work in cities, and the like, it is often necessary to dig trenches and build retaining walls in them before excavating the core of earth between the walls. The following examples of such work are taken from personal records:

Example I.—A Smith mixer was used, the concrete being delivered where wanted by a Lambert cableway of 400 ft. span. The broken stone and sand were delivered near the work in hopper-bottom cars which were dumped through a trestle onto a plank floor. Men loaded the material into one-horse dump carts which hauled it 900 ft. to the mixer platform. This platform was 24×24 ft. square, and 5 ft. high, with a planked approach 40 ft. long and contained 7,300 ft. B. M. The stone and sand were dumped at the mouth of the mixer and shoveled in by 4 men. Eight men, working in pairs, loaded the broken stone into the carts, and 2 men loaded the sand. Each cart was loaded with about 70 shovelfuls of stone on top of which 35 shovelfuls of sand were thrown. It took 3 to 5 minutes to load on the stone and 1 minute to load the sand. The carts traveled very slowly, about 150 ft. a minute—in fact, all the men on the job, including the cart drivers, were slow. After mixing, the concrete was dumped into iron buckets holding 14 cu. ft. water measure, making about ½ cu. yd. in a batch. The buckets were hooked on to the cableway and conveyed where wanted in the wall. Steam for running the mixer was taken from the same boiler that supplied the cableway engine. The average output of this plant was 100 cu. yds. of concrete per 10-hour day, although on many days the output was 125 cu. yds., or 250 batches. The cost of mixing and placing was as follows, on a basis of 100 cu. yds. per day:

	Per day.	Per cu. yd.
8 men loading stone into carts	$12.00	$ .12
2 men loading sand into carts	3.00	.03
1 cart hauling cement	3.00	.03
8 carts hauling stone and sand	24.00	.24
4 men loading mixer	6.00	.06
1 man dumping mixer	1.50	.01
2 men handling buckets at mixer	3.00	.03
6 men dumping buckets and ramming	9.00	.09
12 men making forms at $2.50	30.00	.30
1 cable engineman	3.00	.03
1 fireman	2.00	.02
1 foreman	6.00	.06
1 waterboy	1.00	.01
1 ton coal for cableway and mixer	4.00	.04
Total	$107.50	$1.07

In addition to this cost of $1.07 per cu. yd. there was the cost of moving the whole plant for every 350 ft. of wall. This required 2 days, at a cost of

$100, and as there were about 1,000 cu. yds. of concrete in 350 ft. of wall 16 ft. high, the cost of moving the plant was 10 cts. per cu. yd. of concrete, bringing the total cost of mixing and placing up to $1.17 per cu. yd. As above stated, the whole gang was slow.

The labor cost of making the forms was high, for such simple and heavy work, costing $10 per M. of lumber placed each day. The forms were 2-in. sheeting plank held by 4×6-in. upright studs 2½ ft. apart, which were braced against the sides of the trench. The face of the forms was dressed lumber and all cracks were carefully puttied and sandpapered.

The above costs relate only to the massive part of the wall and not the cost of putting in the facing mortar, which was excessively high. The face mortar was 2 ins. thick, and about 3½ cu. yds. of it were placed each day with a force of 8 men! Two of these men mixed the mortar, 2 men wheeled it in barrows to the wall, 2 men lowered it in buckets, and 2 men put it in place on the face of the wall. If we distribute this labor cost on the face mortar over the 100 cu. yds. of concrete laid each day, we have another 12 cts. per cu. yd.; but a better way is to regard this work as a separate item, and estimate it as square feet of facing work. In that case these 8 men did 500 sq. ft. of facing work per day at a cost of nearly 2-½ cts. per sq. ft. for labor.

Example II.—The building of a wall similar to the one just described was done by another gang as follows: The stone and sand were delivered in flat cars provided with side boards. In a stone car 5 men were kept busy shoveling stone into iron dump buckets having a capacity of 20 cu. ft. water measure. Each bucket was filled about two-thirds full of stone, then it was picked up by a derrick and swung over to the next car which contained sand, where two men filled the remaining third of the bucket with sand. The bucket was then lifted and swung by the derrick over to the platform of the mixer where it was dumped and its contents shoveled by four men into the mixer, cement being added by these men. The mixer was dumped by two men, loading iron buckets holding about ½ cu. yd. of concrete each, which was the size of each batch. A second derrick picked up the concrete bucket and swung it over to a platform where it was dumped by one man; then ten men loaded the concrete into wheelbarrows and wheeled it along a runway to the wall. One man assisted each barrow in dumping into a hopper on the top of a sheet-iron pipe which delivered the concrete. The two derricks were stiff-leg derricks with 40-ft. booms, provided with bull-wheels, and operated by double cylinder (7×10-in.) engines of 18-HP. each. About 1 ton of coal was burned daily under the boiler supplying steam to these two hoisting engines. The output of this plant was 200 batches or 100 cu. yds. of concrete per 10-hr. day, when materials were promptly supplied by the railroad; but delays in delivering cars ran the average output down to 80 cu. yds. per day.

On the basis of 100 cu. yds. daily output, the cost of mixing and placing the concrete was as follows:

	Per day.	Per cu. yd.
5 men loading stone	$ 7.50	$.07½
2 men loading sand	3.00	.03
4 men charging mixer	6.00	.06
2 men loading concrete into buckets	3.00	.03
1 man dumping concrete from buckets	.50	.01½
10 men loading and wheeling concrete	.00	.15
1 man dumping wheelbarrows	1.50	.01½
3 men spreading and ramming	4.50	.04½
2 enginemen	5.00	.05
1 fireman	2.00	.02
1 waterboy	1.00	.01
1 foreman	6.00	.06
10 men making forms	25.00	.25
1 ton coal	4.00	.04
Total	85.00	$.85

In addition there were 8 men engaged in mixing and placing the 2-in. facing of mortar as stated above.

CHICAGO DRAINAGE CANAL.—The method and cost of constructing some 20,000 ft. of concrete wall by contract in building the Chicago

Drainage Canal is compiled from records kept by Mr. James W. Beardsley. The work was done on two separate sections, Section 14 and Section 15. In both cases a 1-1½-4 natural cement concrete was used with a 3-in. facing and a 3-in. coping of 1-3 Portland cement mortar.

Section 14.—The average height of the wall was 10 ft., and the thickness at base was one-half the height. The stone for the concrete was obtained from the spoil bank of the canal, loaded into wheelbarrows and wheeled about 100 ft. to the crusher; some was hauled in wagons. An Austin jaw crusher was used, and it discharged the stone into bins from which it was fed into a Sooysmith mixer. The crusher and the mixer were mounted on a flat car. Bucket elevators were used to raise the stone, sand and cement from their bins to the mixer; the buckets were made of such size as to give the proper proportions of ingredients, as they all traveled at the same speed. Only two laborers were required to look after the elevators. The sand and cement were hauled by teams and dumped into the receiving bins. There were 23,568 cu. yds. on Section 14 and the cost was as follows:

	Typical force.	Wages per 10 hrs.	Cost per cu. yd.
General force:			
Superintendent	1.0	$5.00	$0.026
Blacksmith	1.1	2.75	0.016
Timekeeper	0.5	2.50	0.007
Watchman	0.6	2.00	0.007
Waterboys	3.9	1.00	0.022
Wall force:			
Foreman	0.9	2.50	0.013
Laborers	8.6	1.50	0.073
Tampers	2.3	1.75	0.022
Mixer force:			
Foreman	1.2	2.50	0.017
Enginemen	1.8	2.50	0.025
Laborers	6.7	1.50	0.057
Pump runner	1.0	2.00	0.010

Mixing machines	1.7	1.25	0.012
Timber force:			
Foreman	0.6	2.50	0.008
Carpenters	4.7	2.50	0.057
Laborers	1.2	1.50	0.010
Helpers	5.3	2.50	0.075
Hauling force:			
Laborers	2.6	1.75	0.026
Teams	6.3	3.25	0.116
Crushing force:			
Foreman	0.5	2.50	0.007
Engineman	1.7	2.50	0.023
Laborers	3.5	1.50	0.032
Austin crushers	1.7	1.20	0.011
Loading stone:			
Foreman	1.7	2.50	0.023
Laborers	32.9	1.50	0.280
Total for crushing, mixing and placing			$0.975

The daily costs charged to the mixers and crushers include the cost of coal, at $2 a ton, and the cost of oil.

The gang "loading stone" apparently did a good deal of sledging of large stones, and they also wheeled a large part of it in barrows to the crusher.

The plant cost $9,600, distributed as follows:

2 jaw crushers	$3,000
2 mixers	3,000
Track	1,260
Lumber	500
Pipe	840

Sheds	400
Pumps	600
Total	$9,600

If this first cost of the plant were distributed over the 23,568 cu. yds. of concrete it would amount to 41 cts. per cu. yd.

The cost of the concrete was as follows:

	Per cu. yd.
Utica cement, at $0.65 per bbl.	$0.863
Portland cement, at $2.25 per bbl.	0.305
Sand, at $1.35 per cu. yd.	0.465
Stone and labor, as above given	0.975
Total	$2.608
First cost of plant	$0.407

Section 15.—The conditions on this section were much the same as on Section 14, just described, except that the limestone was quarried from the bed of the canal, and was crushed in a stationary crusher, No. 7 Gates. The stone was hauled 1,000 ft. to the crusher on cars drawn by a cable from a hoisting engine. The output of this crusher averaged 210 cu. yds. per day of 10 hrs. The crushed stone was hauled in dump cars, drawn by a locomotive, to the mixers. Spiral screw mixers mounted on flat cars were used, and they delivered the concrete to belt conveyors which delivered the concrete into the forms.

The forms on Section 15 (and on Section 14 as well) consisted of upright posts set 8 ft. apart and 9 ins. in front of the wall, held at the toe by iron dowels driven into holes in the rock, and held to the rear posts by tie rods. The plank sheeting was made up in panels 2 ft. wide and 16 ft. long, and was held up temporarily by loose rings which passed around the posts which were gripped by the friction of the rings. These panels were brought to proper line and held in place by wooden wedges. After the concrete had set 24 hrs. the wedges were struck, the panels removed and scraped clean ready to be used again.

The cost of quarrying and crushing the stone, and mixing the concrete on Section 15 was as follows:

	Typical force.	Wages per 10 hrs.	Cost per cu. yd.
General force—			
Superintendent	1.0	$5.00	$0.024
Blacksmith	0.9	2.75	0.011
Teams	1.7	3.00	0.025
Waterboy	4.5	1.00	0.022
Wall force—			
Foreman	1.1	2.50	0.010
Laborers	14.4	1.50	0.105
Tampers	0.1	1.75	0.001
Mixer force—			
Foreman	2.1	2.50	0.026
Enginemen	2.1	2.50	0.022
Laborers	23.1	1.50	0.180
Mixing machines	2.1	1.25	0.022
Timber force—			
Carpenters	0.8	3.00	0.013
Laborers	0.7	1.50	0.005
Helpers	10.2	2.50	0.125
Hauling force—			
Foreman	0.7	2.50	0.009
Enginemen	1.4	2.50	0.019
Fireman	0.4	1.75	0.003
Brakeman	2.2	2.00	0.018
Teams	0.4	3.25	0.007
Laborers	1.5	1.50	0.010
Locomotives	1.4	2.25	0.015

Crushing force—			
Foreman	1.0	2.50	0.014
Enginemen	1.0	2.50	0.014
Laborers	11.1	1.50	0.081
Firemen	1.0	1.75	0.008
Gyratory crusher	1.0	2.25	0.011
Quarry force—			
Foreman	1.2	2.50	0.012
Laborers	19.0	1.50	0.140
Drillers	1.8	2.00	0.017
Drill helpers	1.8	1.50	0.013
Machine drills	1.8	1.25	0.011
Total			$0.993

The first cost of the plant for this work on Section 15 was $25,420, distributed as follows:

1 crusher, No. 7 Gates	$12,000
Use of locomotive	2,200
Car and track	5,300
3 mixers	3,000
Lumber	1,200
Pipe	720
Small tools	1,000
Total	$25,420

This $25,420 distributed over the 44,811 cu. yds. of concrete amounts to 57 cts. per cu. yd.

It will be noted that 2 mixers were kept busy. Their average output was 100 cu. yds. each per day, which is the same as for the mixers on Section 14.

The total cost of concrete on Section 15 was as follows:

	Per cu. yd.
Labor quarrying, crushing and mixing	$0.991
Explosives	0.083
Utica cement, at $0.60 per bbl.	0.930
Portland cement, at $2.25 per bbl.	0.180
Sand, at $1.35 per cu. yd.	0.476
Total	$2,660
First cost of plant	$0.567

It is not strictly correct to charge the full first cost of the plant to the work as it possessed considerable salvage value at the end.

Comparison.—For the purpose of comparing Sections 14 and 15 the following summary is given of the cost per cubic yard of concrete:

	Sec. 14.	Sec. 15.
General force	$0.078	$0.082
Wall force	0.108	0.116
Mixing force	0.121	0.250
Timbering force	0.150	0.140
Hauling force	0.142	0.081
Crushing force	0.073	0.128
Quarry force	0.303	0.275
Cement, natural	0.863	0.930
Cement, Portland	0.305	0.180
Sand	0.465	0.476
Plant (full cost)	0.407	0.567
Total	$3.015	$3.225

It should be remembered that on Section 14 there was no drilling and blasting of the rock, but that the "quarry force" not only loaded but hauled the stone to the crusher. The cost of mixing on Section 15 is higher than on

Section 14 because the materials were dumped on platforms and shoveled into the mixer, instead of being discharged from bins into the mixer as on Section 14.

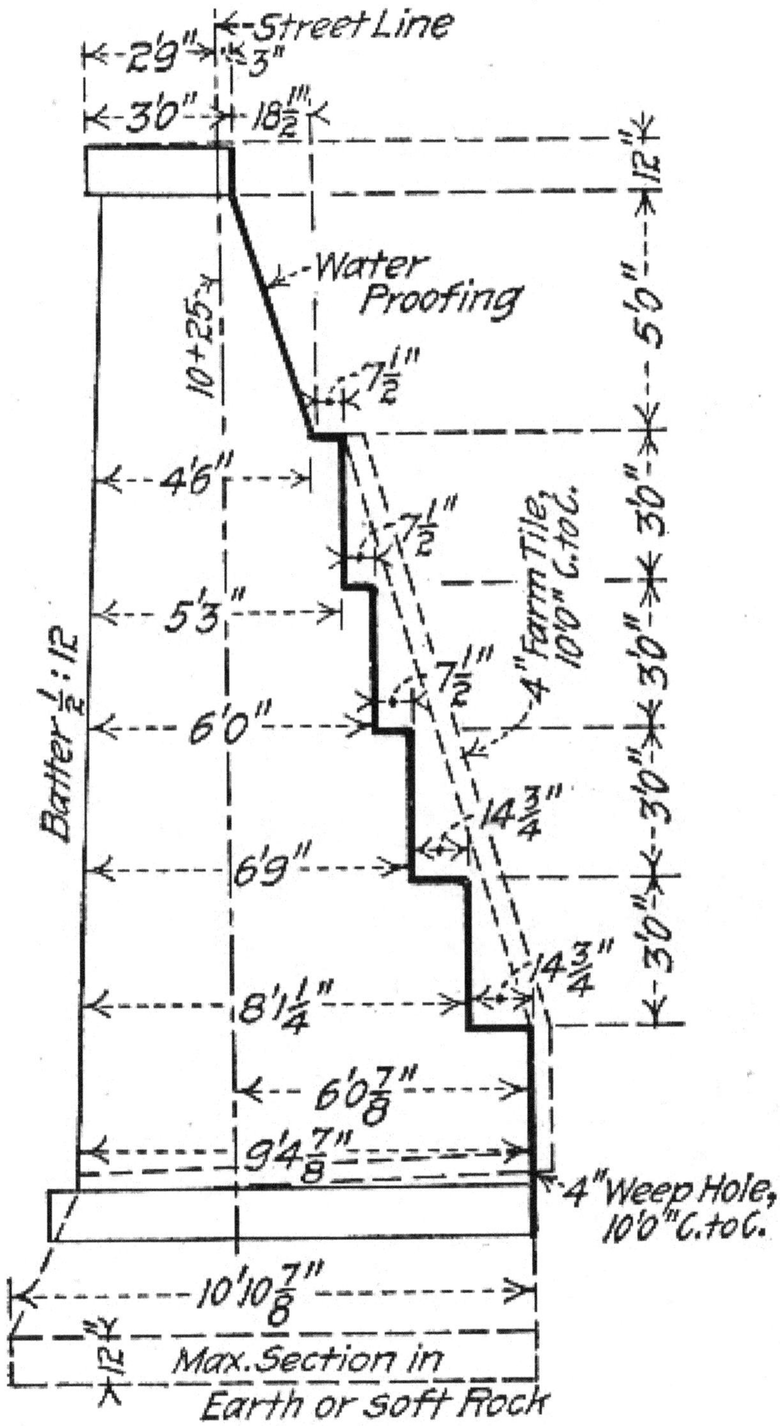

Fig. 108.—Cross Section of Retaining Wall.
New York Central Terminal Work.

Fig. 109.—Portable Concrete Mixing Tower, N. Y. Central Terminal Work.

GRAND CENTRAL TERMINAL, NEW YORK, N. Y.—In building a retaining wall of the cross-section, shown in Fig. 108, a traveling tower moving on tracks parallel to the wall contained the concrete mixing plant. The construction of the tower is shown in Fig. 109. The tower had two platforms, one at the top carrying two 10-cu. yd. bins for sand and stone and the other directly below carrying 40 cu. ft. (4 cu. ft. cement, 12 cu. ft. sand and 24 cu. ft. stone) Ransome mixer driven by a 30 H.P. motor and a Lidgerwood motor hoist.

The elevator tower carried two 40-cu. ft. Ransome dumping buckets traveling in guides and dumping automatically into the bins. These buckets were operated by the Lidgerwood motor hoist on the mixer platform. Sand and broken stone on flat cars were brought alongside the tower. The sand was shoveled direct from the car into the sand bucket, but the broken stone was shoveled into wheelbarrows which were wheeled over a light bridging from car to bucket and dumped.

Wheelbarrows were used for handling the stone chiefly because the capacity of the plant was so great that enough men could not be worked in the limited space around the bucket to keep up the supply by shoveling. The wheelbarrow work added materially to the cost. Cement was carried from the cars to the sand bucket, hoisted and stored on the mixer platform which provided storage room for 100 bags. A 1-3-6 mixture was used; the sand and stone were chuted directly from the bins to the charging hopper and the cement was charged by hand.

The mixed concrete was delivered to two 1 cu. yd. dump cars running on a 2-ft. gage track laid in sections on thecross pieces connecting the uprights of the forms. The track had no switches, so that one car had to wait for the other. Four men were required to push each car and two more men assisted in dumping the car and kept the track clear. The wall was built in sections 51 ft. long, each containing 250 cu. yds.

One of these sections was filled in 8 hours with ease and by a little hustling a section was filled in 6¾ hours, which is at the rate of 37 cu. yds. of concrete per hour. Working 8 hours per day the cost of mixing, transporting and placing concrete with this mixing plant, with wages for common labor of $1.50 per day, was as follows:

	Total.	Per cu. yd.
2 men carrying cement	$3.00	$0.012
6 men shoveling sand	9.00	0.036
17 men shoveling stone	25.00	0.100
11 men wheeling stone	16.00	0.064
2 men at stone and sand bins	3.00	0.012
2 men opening cement bags	3.00	0.012
1 man dumping hopper	1.50	0.006
1 man dumping mixer	1.50	0.006
1 man cleaning chute, mixer, etc.	1.50	0.006
1 motorman or engineer	3.00	0.012
Total labor mixing	$66.50	$0.266
8 men pushing 2 cars	12.00	0.048
2 men cleaning track, etc.	3.00	0.012
7 men spading concrete	10.50	0.042
Total labor transporting, placing	$ 25.50	$0.102
1 foreman	5.00	0.020
Electricity estimated	7.00	0.028
Total general	$ 12.00	$0.048
Grand total	$104.00	$0.416

It will be noted that the cost of shoveling and wheeling the broken stone amounts to 16.4 cts. per cu. yd., or nearly 40 per cent. of the total cost of mixing and placing. The cost of spading the concrete is also high for a sloppy mixture, but is probably accounted for by the fact that the concrete had to be spaded so as to have 2 or 3 ins. of clear mortar next the forms. The forms used in constructing the wall are shown by Figs. 110 and 111. They were made in panels 51 ft. long and a locomotive crane was used to shift the panels. This crane worked handling forms only a small part of the time, but a form gang of 10 carpenters was kept busy all of the time moving and reassembling. Assuming the work of the crane to amount to $5 per day and the wages of the carpenter gang to amount to $25, we get a cost of 12 cts. per cubic yard of concrete for shifting forms. It should be noted carefully that the costs given for this work do not include cost of materials, interest on plant, superintendence and other items.

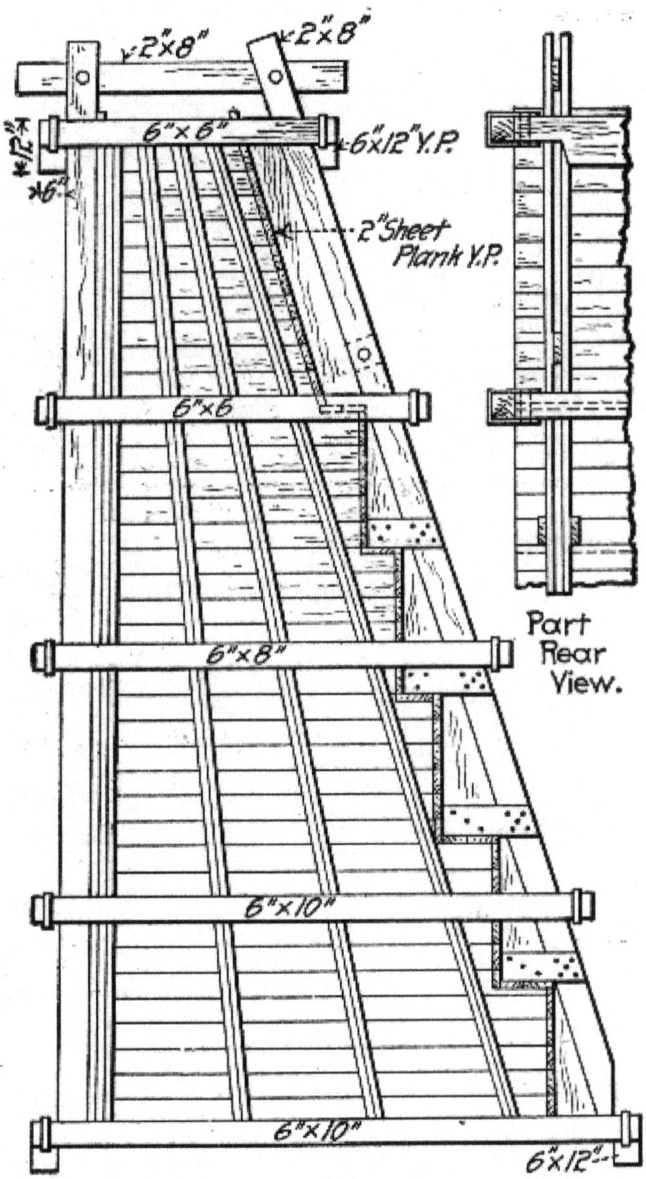

Fig. 110.—End View of Forms for Retaining Wall, New York Central Terminal Work.

WALL FOR RAILWAY YARD.—For building a retaining wall 7 ft. high, forms were made and placed by a carpenter and helper at $8 per M., wages being 35 cts. and 20 cts. an hour, respectively. Concrete materials were dumped from wagons alongside the mixing board. Ramming was unusually thorough. Foreman expense was high, due to small number in gang; 2 cu. yds. were laid per hour by the gang.

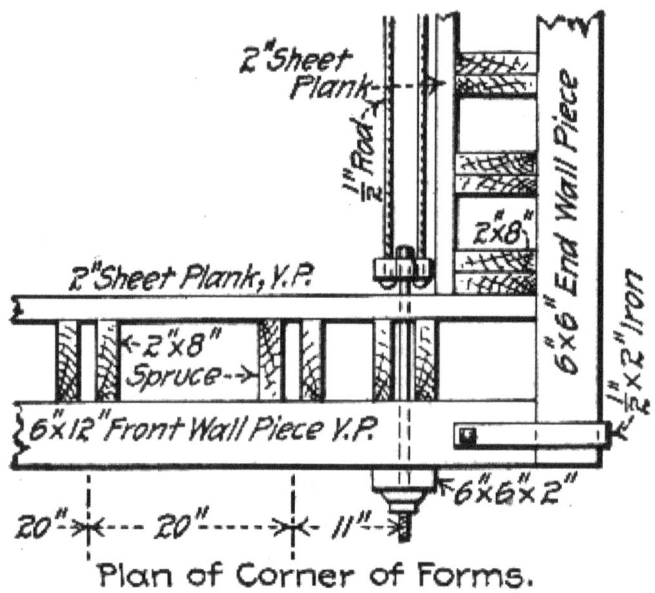

Fig. 111.—Corner Detail of Retaining Wall Forms, New York Central Terminal Work.

	Per day.	Per cu. yd.
7 mixers, 15 cts. per hour	$10.50	$0.53
2 rammers, 15 cts. per hour	3.00	0.15
1 foreman 30 cts. per hr., 1 waterboy 5 cts.	3.50	0.17
Total labor	$17.00	$0.85

The total cost was as follows per cubic yard:

	Per cu. yd.
0.8 bbls. Portland cement, at $2	$1.60
Sand	0.30
Gravel	0.70
Labor mixing and placing	0.85
Lumber for forms, at $16 per M.	0.56
Labor on forms, at $8 per M.	0.28

Total, per cubic yard	$4.29

The sheathing plank for the forms was 2-in. hemlock.

CONCRETE FOOTING FOR RUBBLE MASONRY RETAINING WALL.—In constructing a footing for a retaining wall at Grand Rapids, Mich., a 1-2½-5 natural cement concrete was used. It was found that 1 cu. yd. of concrete was equivalent to 29.8 cu. ft. of material composed of 3.6 cu. ft. or 1.1 bbls. of cement, 8.4 cu. ft. or 2.7 bbls. of sand and 17.8 cu. ft. or 5.5 bbl. of broken stone. The labor cost of 15.5 cu. yds. of concrete was as follows:

Item.	Total.	Per cu. yd.
Foreman, 14 hours at 40 cts.	$ 5.60	$0.3613
Foreman, 20 hours at 22.5 cts.	4.50	0.2903
Laborers, 49 hours at 12.5 cts.	6.11	0.3942
Mason, 2 hours at 35 cts.	0.70	0.0451
Total labor	$16.91	$1.0909

All material was furnished by the railway company, the contractor furnishing labor only; his contract price for this was $1 per cu. yd.

TRACK ELEVATION, ALLEGHENY, PA.—The wall was 6,100 ft. long and 75 per cent. was on curves. The first wall built had a top width of 2½ ft. and a bottom width of 0.4 the height with the back on a smooth batter. Later the back was stepped and last the wall was proportioned as follows: Calling the height from top of foundation to under coping, then width of base was 0.45 (h + 3), the top measuring 2½ ft. The back was arranged in steps 24 ins., 30 ins. and 36 ins. high, and the thickness of wall at each step was, calling h equal to height of step from base, 0.45 (h + 3). Several forms of expansion joints were tried. The first was tarred paper extending through the wall every 50 ft.; the second was ½-in. boards running through the wall every 50 ft.; the third was ½-in. board extending 2 ft. into the wall, with a ¼-in. cove at the angles, every 25 ft. The third construction gave perfect satisfaction.

A 1-2-5 natural cement and a 1-3-6 Portland cement concrete mixed fairly wet were used. The concrete was laid in 8-in. courses and faced with a 1-2 mortar. The forms were 2-in. white pine faced and jack planed on the edges; upon removal of the forms board marks and other defects were removed and a wash of neat cement was applied. One contractor used hand mixing. The sand and gravel were measured in wheelbarrows and wheeled onto the platform; the sand and cement were spread in thin layers, one over the other, and thoroughly mixed dry; the gravel was then spread over the mixture, the whole was shoveled into barrows or the pit again shoveled into place and rammed. The other contractor used a cubical mixer. A charging box holding 1¼ cu. yds. and graduated to show the correct proportions of sand and gravel was filled by shoveling; cement was placed on top and the box hoisted and dumped into the mixer. A barrel holding the correct amount of water was emptied into the mixer which was turned 10 or 15 times and discharged into cars. The costs of mixing by hand and by machine were as follows:

Hand mixing.	Total.	Per cu. yd.
½ foreman at $3	$ 1.50	$0.025
3 men wheeling barrows at $1.50	4.50	0.075
10 men wheeling materials at $1.50	15.00	0.250
3 men mixing sand and gravel at $1.50	4.50	0.075
6 men mixing concrete at $1.50	9.00	0.150
1 man sprinkling at $1.50	1.50	0.025
Total	$36.00	$0.600

The output of the hand mixing gang was 60 cu. yds. per day.

Machine mixing.	Total.	Per cu. yd.
1 foreman at $3.50	$ 3.50	$0.035
1 stationary engineer at $3	3.00	0.030
½ foreman at $1.75	0.87	0.009
15 men loading charging bucket at $1.50	22.50	0.225
2 men dumping charging bucket at $1.75	3.50	0.035

2 tagmen at $2, ½ time	2.00	0.020
1 man at trap at $2, ½ time	1.00	0.010
Total	$36.37	$0.364

The output of the cubical mixer was 100 cu. yds. per day.

The costs of placing concrete in the forms above the foundation by hand below 12 ft., and by cars and derricks any height, were as follows:

By hand (barrows) below 12 ft.	Total.	Per cu. yd.
4 men loading concrete at $1.50	$ 6.00	$0.100
1 foreman ½ time at $3	1.50	0.025
10 men wheeling at $1.50	15.00	0.250
1 man scraping barrows at $1.50	1.50	0.025
2 men placing concrete at $1.50	3.00	0.050
1 man placing mortar face at $1.50	1.50	0.025
2 men mixing and carrying mortar at $1.50	3.00	0.050
Total	$31.50	$0.525
By cars and derricks—		
1 horse and driver at $3	$ 3.00	$0.030
2 men dumping concrete ½ time at $1.50	1.50	0.015
1 fireman ½ time at $1.75	0.88	0.009
3 tagmen at $1.50	4.50	0.045
8 men placing and ramming conc. at $1.50	12.00	0.120
2 men mixing mortar at $1.50	3.00	0.030
2 men placing mortar at $1.50	3.00	0.030
2 men carrying mortar at $1.50	3.00	0.030
1 foreman at $3	3.00	0.030
1 stationary engineer at $3	3.00	0.030
2 men attending hook at $1.50	3.00	0.030

Total	$39.88	$0.399

The costs of placing concrete in the foundations were as follows:

By hand—	Total.	Per cu. yd.
1 foreman ½ time at $3	$ 1.50	$0.025
4 men shoveling concrete at $1.50	6.00	0.100
1 man placing concrete at $1.50	1.50	0.025
1 man ramming concrete at $1.50	1.50	0.025
Total	$10.50	$0.175
By machine—		
1 horse and driver at $3	$ 3.00	$0.030
3 men pushing and unloading car at $1.50	4.50	0.045
5 men placing and ramming at $1.50	7.50	0.075
1 foreman at $3	3.00	0.030
2 men dumping mixer at $1.50	3.00	0.030
Total	$21.00	$0.210

COST OF RETAINING WALL.—The following figures of the cost of a concrete retaining wall are given by C. C. Williams:

Cost of Material.

Kind and amount of material—	Unit Price.	Cost.
Stone, 441 tons	$ .70	$308.70
Sand, 182.5 yds.	.55	100.37
Cement, 536 bbls.	.85	453.60
Total		$862.67
Lumber ¾ value		$205.33
Wheelbarrows, ¾ value, 6 at $3.50		15.75
Total		$221.08

Excavation—	
Labor, 4,002 hours at 15 cts.	$600.30
Carts, 800 hours at 12½ cts.	100.00
Foreman, 460 hours at 35 cts.	171.00
Waterboy, 240 hours at 10 cts.	24.00
Total	$895.30
Concrete—	
Labor, 2,398 hours at 15 cts.	$359.70
Foreman, 224 hours at 35 cts.	77.40
Total	$437.10
Handling material—	
Unloading cars, 380 hours at 15 cts.	$ 57.00
Foreman, 40 hours at 35 cts.	14.00
Total	$ 71.00
Forms—	
Carpenters, 997 hours at 22½ cts.	$224.33
Work to support bridge—	
Carpenters, 542 hours at 22½ cts.	$121.95
Labor, 458 hours at 15 cts.	68.70
Total	$190.65
Superintendence and office—	
Superintendent, 30 hours at 50 cts.	$15.00
Office	20.00
Total	$35.00
Grand total	$2,937.13

Proportional costs—

Item.	Cost.	Cost Per Yard of Concrete.	Per Cent. of Total Cost.
Concrete materials	$ 862.67	$2.02	46.7
Laying concrete	437.10	1.03	23.4
Lumber	205.33	.48	11.3
Building forms	224.33	.53	12.3
Handling material	71.00	.17	03.8
Wheelbarrows	15.75	.04	01.0
Supt., etc.	35.00	.07	01.5
Total	$1,851.18	$4.34	100.00
Work on bridge	190.65		
Excavation	895.30		
	$2,937.13		

14

CONCRETE PAVEMENT FOUNDATIONS

Contractor's skill or want of skill in systematizing and managing labor counts as high in street work as in any class of concrete construction. As previously demonstrated, the cost of mixing is a very small portion of the labor cost of concrete in place; the costs of getting the materials to the mixer and the mixed concrete to the work are the big items, and in street work the opportunity for increasing the cost of these items through mismanagement is magnified by the large area of operations involved per cubic yard of concrete placed. One cubic yard of concrete makes 6 sq. yds. of 6-in. pavement foundations and 100 cu. yds. of concrete make a 6-in. foundation for 300 ft. of 30-ft. street, while 4 to 5 cu. yds. will build 100 ft. of ordinary curb and gutter. Thus the haulage per cubic yard is considerable at best, and lack of plan in distributing stock piles and handling the concrete can easily result in such increased haulage expenses as to change a possible profit into a certain loss. A little thought and skill in planning street work pays a good profit.

MIXTURES EMPLOYED.—A comparatively lean concrete will serve for pavement foundations; mixtures of 1-4-8 Portland cement or 1-2-5 natural cement are amply good and it is folly, ordinarily, to employ richer mixtures. Until recently, natural cement has been used almost exclusively; a 1-2-5 natural cement mixture requires about 1.15 bbls. of cement per cubic yard of concrete. A 1-4-8 Portland cement mixture requires about 0.7 bbl. of cement per cubic yard. In the opinion of the authors a considerably leaner mixture of Portland concrete is sufficiently good when it is well mixed in machine mixers—for a 6-in., foundation 0.5 bbl. per cu. yd. The mixtures actually employed are proportioned about as stated and their cost, or that of any other common mixture, may easily be computed from Tables XII and XIII, giving for different mixtures the quantities of cement, sand and stone per cubic yard of concrete; the product of these quantities and the local prices of materials in the stock piles gives the cost. When the concrete is mixed by hand the ordinary labor cost of foundations is 0.4 to 0.5 of a 10-hour day's wages per cubic yard of concrete; occasionally it may be as low as 0.3 of a day's wages where two mixing gangs are worked side by side under different foremen and with an exacting contractor. Data for machine mixing are too few to permit a similar general statement for machine work, but in one case coming under the authors' observation, the cost figured out to a little less than 0.2 of a day's wages per cubic yard.

DISTRIBUTION OF STOCK PILES.—Assuming a 30-ft. street and a 1-3-5 concrete laid 6 ins. thick, the quantities of concrete materials required per lineal foot of street are: Cement 0.60 bbl., sand 0.27 cu. yd., stone 0.44 cu. yd. The stock piles should be so distributed that each supplies enough materials for a section of foundation reaching half way to the next adjacent stock pile on each side, and they should not contain more or less material, otherwise a surplus remains to be cleaned up or a deficiency to be supplied by borrowing from another pile. A little care will ensure the proper distribution and it is well paid for in money saved by not rehandling surplus or borrowed materials. For a given mixture and a given width and thickness of foundation, the sizes of the stock piles are determined by their distance apart and this will depend upon whether hand or machine mixing is employed and upon the means adopted for hauling the raw materials and the mixed concrete. It is worth while always in stock piles of any size, to lay a flooring of plank particularly under the stone pile; if dumped directly on the ground it costs half as much again to handle stone. Current practice warrants everything from a continuous bank, to piles from 1,000 to 1,500 ft. apart, in the spacing of stock piles.

HINTS ON HAND MIXING.—All but a small percentage of the concrete annually laid in street work is hand mixed. The authors are confident that this condition will disappear as contractors learn more of the advantages of machine mixing, but it prevails at present. The general economics of hand mixing are discussed in Chapter II; in street work as before stated, the big items of labor cost are the costs of handling materials and the data in Chapter II on these processes deserve special attention. It is particularly worth noting that it is seldom economical to handle materials in shovels where carrying is necessary; it is a common thing in street work to see an attempt to get the stock piles so close to the mixing board that the material can be handled with shovels, and this is nearly always an economic error. Street work is readily measured; in fact, its progress can be seen at a glance, and advantage can often be taken of this fact to profit by the rivalry of separate gangs. The authors have known of the labor costs being reduced as much as 25 per cent., due to pitting one gang against another where each could see the progress made by the other.

METHODS OF MACHINE MIXING.—Concrete mixers have been slow to replace handwork in laying pavement foundations. In explanation of this fact it is asserted: (1) That frequent shifting of the mixer causes too much lost time, and (2) that the principal item of labor cost in street work is the conveying of materials to and from the mixer, and this item is the same whether hand or machine mixing be employed. The records of machine mixer work given elsewhere in this chapter go far, in the opinion of the authors, toward disproving the accuracy of both assertions. If the machine used and the methods of work employed are adapted to the conditions of street work, machine mixing can be employed to decided advantage.

A continuous and large output is demanded in a mixer for street work; the perfection of the mixing is within limits a minor consideration. This at once admits for consideration types of mixers whose product is classed as unsuitable for reinforced concrete work, and also admits of speeding up the output of the better types to a point beyond that at which they turn out their most perfect product. Keeping these facts in mind either of the following two systems of work may be employed: (1) Traction plants which travel with the work and deposit concrete in place, or so nearly in place that little shoveling is necessary; (2) portable plants which are set up at wide intervals along the work and which discharge the concrete into carts or dump wagons which distribute it to the work.

The secret of economic work with plants of the class cited first is the distribution of the stock piles so as practically to eliminate haulage from stock pile to mixer. The mixer backs away from the work, its discharge end being toward the work and its charging end away from it. Then deposit the materials so as to form a continuous stock pile along the center of the street; the mixer moving backward from the completed foundation keeps close to the materials and if the latter are uniformly distributed in the pile the great bulk of the charging is done by shoveling direct into the charging bucket.

The point to be watched here is that the shovelers do not have to carry the materials; separate stock piles within moderate hauling distance by wheelbarrows are a far more economic arrangement than a continuous pile so irregularly distributed that much of the material has to be carried even a few paces in shovels.

Economic work with plants of the second class depends upon efficient and adequate means of hauling the mixed concrete to the work. The plant should not be shifted oftener than once in 1,000 to 2,000 ft., or, say, four city blocks. This does away with the possibility of wheelbarrow haulage; large capacity hand or horse carts must be employed. With 6 cu. ft. hand carts, such as the Ransome cart, a haul of 500 ft. each way from the mixer is possible and with horse carts, such as the Briggs, this economic distance is increased to 1,000 ft. each way from the mixer. The mixer must be close to the stock pile and it will pay to make use of improved charging devices. A 6-in. foundation for 2,000 ft. of 30-ft. street calls for 667 cu. yds. of concrete, and if both sides are curbed at the same time, 100 cu. yds. more are added, or 767 cu. yds. in all; where intersecting streets are to be paved in both directions from the mixer plant these amounts are doubled. A very small saving per cubic yard due to mechanical handling of the materials to the mixer amounts to the interest on a considerable investment in such plant. A point that should not be forgotten is that carts such as those named above spread the concrete in dumping so that little or no shoveling is required.

FOUNDATION FOR STONE BLOCK PAVEMENT, NEW YORK, N. Y.—Mr. G. W. Tillson, in "Street Pavements and Paving Materials," p. 204, gives the following data on the cost of granite block pavement in New York City in 1899. The day was 10 hours long:

Concrete gang—	Per day.	Per sq. yd.	Per cu. yd.
1 foreman	$ 3.00	$0.0125	$0.075
8 mixers on two boards, at $1.25	10.00	0.0416	0.250
4 wheeling stone and sand, at $1.25.	5.00	0.0208	0.125
1 carrying cement and supplying water, at $1.25	1.25	0.0051	0.031

1 ramming, at $1.25	1.25	0.0051	0.031
Total, 240 sq. yds. (40 cu. yds.).	$20.50	$0.0851	$0.512

The concrete was shoveled direct from the mixing boards to place.

Cost 1-2-4 concrete—	Per cu. yd.
1⅓ bbls. natural cement, at $0.90	$1.20
0.95 cu. yd. stone, at $1.25	1.19
0.37 cu. yd. sand, at $1.00	0.37
Labor	0.51
	$3.27

In laying 5,167 sq. yds. of granite block pavement on one job in New York City in 1905, the authors' records show that one laborer mixed and laid 1.3 cu. yds. of concrete per day in a 6-in. foundation; this is a very small output. The work was done by contract and the labor cost was as follows:

Item.	Total.	Per sq. yd.	Per cu. yd.
28½ days foreman at $3.50	$ 99.75	$0.0193	$0.118
399 days laborers at $1.75	698.25	0.1351	0.826
	$798.00	$0.1544	$0.944

The average day's wages was $1.86, so that the labor cost was about 0.5 of a day's wages per cubic yard of concrete.

FOUNDATION FOR PAVEMENT, NEW ORLEANS. LA.—Mr. Alfred E. Harley states that in laying concrete foundations for street pavement in New Orleans, a day's work, in running three mixing boards, covering the full width of the street, averaged 900 sq. yds., 6 ins. thick, or 150 cu. yds., with a gang of 40 men. With wages assumed to be 15 cts. per hour the labor cost was:

Cts. per cu. yd.

6 men wheeling broken stone	6
3 men wheeling sand	3
1 man wheeling cement	1
2 men opening cement	2
7 men dry mixing	7
8 men taking concrete off	8
3 men tamping	3
3 men grading concrete	3
1 man attending run planks	1
3 water boys	1
2 extra men and 1 foreman	4
Total labor cost	39 cts.

FOUNDATIONS FOR STREET PAVEMENT, TORONTO, CANADA.—The following cost of a concrete base for pavements at Toronto has been abstracted from a report (1892) of the City Engineer, Mr. Granville C. Cunningham. The concrete was 1-2½-7½ Portland; 2,430 cu. yds. were laid, the thickness being 6 ins., at the following cost per cubic yard:

0.77 bbl. cement, at $2.78	$2.14
0.76 cu. yd. stone, at $1.91	1.45
0.27 cu. yd. sand and gravel, at $0.80	0.22
Labor (15 cts. per hr)	1.03
Total	$4.84

Judging by the low percentage of stone in so lean a mixture as the above, the concrete was not fully 6 ins. thick as assumed by Mr. Cunningham. Note that the labor cost was 1½ to 2 times what it would have been under a good contractor.

MISCELLANEOUS EXAMPLES OF PAVEMENT FOUNDATION WORK.—The following records of pavement foundation work are taken from the note and time books of one of the authors:

Case 1.—Laying 6-in. pavement foundation; stone delivered and dumped upon 2-in. plank laid to receive it. Sand and stone were dumped along the street, so that the haul in wheelbarrows to mixing board Was about 40 ft. Two gangs of men worked under separate foremen, and each gang averaged 4.5 cu. yds. concrete per hour. The labor cost was as follows for 45 cu. yds. per gang:

	Per day.	Per cu. yd.
4 men filling barrows with stone and sand ready for the mixers, wages 15 cts. per hour	$6.00	$0.13
10 men, wheeling, mixing and shoveling to place (3 or 4 steps), wages 15 cts. per hour	15.00	0.33
2 men ramming, wages 15 cts. per hour	3.00	0.07
1 foreman at 30 cts. per hour and 1 water boy, 5 cts	3.50	0.08
Total	$27.50	$0.61

Case II.—Sometimes it is desirable to know every minute detail cost, for which purpose the following is given:

	Per cu. yd. Day's labor.	Cost.
3 men loading stones into barrows	$0.06	$0.09
1 man loading sand into barrows	0.02	0.03
2 men ramming	0.04	0.06
1 foreman and 1 water boy equivalent to	0.035	0.05
Wheeling sand and cement to mixing board	0.02	0.03
Wheeling stone to mixing board	0.026	0.04
9 men mixing mortar	0.013	0.02
Mixing stone and mortar	0.049	0.07
Placing concrete (walking 15 ft.)	0.072	0.11
Total	$0.335	$0.50

In one respect this is not a perfectly fair example (although it represents ordinary practice), for the mortar was only turned over once in mixing instead of three times, and the stone was turned only twice instead of three or four times. Water was used in great abundance, and by its puddling action probably secured a very fair mixture of cement and sand, and in that way secured a better mixture than would be expected from the small amount of labor expended in actual mixing. About 9 cts. more per cu. yd. spent in mixing would have secured a perfect concrete without trusting to the water.

Case III.—Two gangs (34 men) working under separate foremen averaged 600 sq. yds., or 100 cu. yds. of concrete per 10-hour day for a season. This is equivalent to 3 cu. yds. per man per day. The stone and sand were wheeled to the mixing board in barrows, mixed and shoveled to place. Each gang was organized as follows:

	Per day.	Per cu. yd.
4 men loading barrows	$ 6.00	$0.12
9 men mixing and placing	13.50	0.27
2 men tamping	3.00	0.06
1 foreman	2.50	0.05
Total	$25.00	$0.50

These men worked with great rapidity. The above cost of 50 cts. per cu. yd. is about as low as any contractor can reasonably expect to mix and place concrete by hand in pavement work.

Case IV.—Two gangs of men, 34 in all, working side by side on separate mixing boards, averaged 720 sq. yds., or 120 cu. yds., per 10-hour day. Each gang was organized as follows:

	Per day.	Per cu. yd.
6 men loading and wheeling	$ 9.00	$0.15
8 men mixing and placing	12.00	0.20
2 men tamping	3.00	0.05

1 foreman	3.00	0.05
Total	$27.00	$0.45

Instead of shoveling the concrete from the mixing board into place, the mixers loaded it into barrows and wheeled it to place. The men worked with great rapidity.

Mr. Irving E. Howe gives the cost of a 6-in. foundation of 1-3-5 natural cement at Minneapolis, Minn., in 1897, as $2.80 per cu. yd., or $0.467 per sq. yd. Cement cost 76 cts. per barrel and stone and sand cost delivered $1.15 and 30 cts. respectively. Mixers received $1.75 per day.

Mr. Niles Meriwether gives the cost of materials and labor for an 8-in. foundation constructed by day labor (probably colored) at Memphis, Tenn., in 1893, as follows:

	Per sq. yd.
Natural cement at $0.74 per bbl	$0.195
Sand at $1.25 per cu. yd	0.075
Stone at $1.87 per cu. yd	0.355
Labor mixing and placing	0.155
Total	$0.780

Labor was paid $1.25 to $1.50 per 8-hour day and 1.16 bbls. of cement were used per cubic yard of concrete. The cost of materials, as will be noted, was high and the labor seems to have been inefficient.

FOUNDATIONS FOR BRICK PAVEMENT, CHAMPAIGN, ILL.— The concrete foundation for a brick pavement constructed in 1903 was 6 ins. thick; the concrete used was composed of 1 part natural cement, 3 parts of sand and gravel, and 3 parts of broken stone. All the materials were mixed with shovels, and were thrown into place from the board upon which the mixing was done. The material was brought to the steel mixing board in wheelbarrows from piles where it had been placed in the middle of the street, the length of haul being usually from 30 to 60 ft. The foundation was 6 ins. thick and it cost as follows for materials and labor:

	Cost per cu. yd.
1.2 bbls. cement, at $0.50	$0.600
0.6 cu. yd. sand and gravel, at $1	0.600
0.6 cu. yd. broken stone, at $1.40	0.840
6 men turning with shovels, at $2	0.080
4 men throwing into place, at $2	0.053
2 men handling cement, at $1.75	0.023
1 man wetting with hose, at $1.75	0.012
2 men tamping, at $1.75	0.023
1 man leveling, at $1.75	0.012
6 men wheeling stone, at $1.75	0.070
4 men wheeling gravel, at $1.75	0.047
1 foreman, at $4	0.027
	$2.387

This is practically 40 cts. per sq. yd., or $2.40 per cu. yd. of concrete for materials and labor. It is evident from the above quantities that a cement barrel was assumed to hold about 4.5 cu. ft., hence the cement was measured loose in making the 1-3-3 concrete.

The accuracy of the quantities given is open to serious doubt. It will also be noted that the labor cost of making and placing the concrete was only 35 cts. per cu. yd., wages being nearly $1.85 per day. This is so remarkably low that some mistake would seem to have been made in the measurement of the work. The authors do not hesitate to say that no gang of men ever made any considerable amount of concrete by hand at the rate of 5.75 cu. yds. per man per day.

Fig. 112.—Foote Continuous Mixer Arranged for Pavement Foundation Work.

FOUNDATION CONSTRUCTION USING CONTINUOUS MIXERS. —The following are records of two jobs of pavement foundation work using continuous mixers with one-horse concrete carts in one instance and wheelbarrows in the other instance. The mixer used was the Foote mixer, as arranged for the work being described it is shown by Fig. 112. One particular advantage of this and similar mixers for street work is that no proportioning or measuring of the materials is required of the men. The mixers are provided with an automatic measuring device, by means of which any desired proportion of cement, sand and stone is delivered to the mixing trough.

The mixer is mounted on trucks, and the hoppers that receive the sand and stone are comparatively low down. The sand can be wheeled in barrows up a run plank and dumped into a hopper on one side of the mixer, and in like manner the gravel or broken stone can be delivered into a hopper on the other side. The cement is delivered in bags or buckets to a man who dumps it into a cement hopper directly over the mixer. All that the operator needs to attend to is to see that the men keep the hoppers comparatively full. The records of work on the two jobs mentioned are as follows:

Fig: 113.—Briggs Cart Distributing Concrete for Pavement Foundation.

Job I.—The sand was delivered from the stock pile by a team hitched to a drag scraper, and was dumped alongside the mixer where two men shoveled it into the hopper. On the same job the concrete was hauled away from the mixer in Briggs' concrete carts. With a gang of 30 men and 2 to 4 horses hauling concrete in Briggs' carts, the contractor averaged 1,200 sq. yds., or 200 cu. yds., per day of 10 hours. With wages of laborers at 15 cts. per hour, and a single horse at the same rate, the cost of labor was 26 cts. per cu. yd., or less than 4½ cts. per sq. yd. of concrete base 6 ins. thick. The coal was a nominal item, and did not add 1 ct. per cu. yd. to the cost. In this case the mixer was set up on a side street and the concrete was hauled in the carts for a distance of a block each way from the mixer. At first four carts were used, but as the concreting approached the mixer, less hauling was required, and finally only two carts were used. An illustration of a Briggs cart is given by Fig. 113; it is hauled by one horse, which the driver leads, and is dumped by an ingenious device operated from the horse's head. The cart dumps from the bottom and spreads the load in a layer about 8 or 9 ins.

thick, so that no greater amount of shoveling is necessary than when barrows are used. It took about 20 seconds for the cart to back up and get its load and about 5 seconds to dump and spread the load.

Job II.—In this job the mixer was charged with wheelbarrows and wheelbarrows were also employed to take the mixed concrete to the work, the mixer being moved forward at frequent intervals. The stock piles were continuous, sand on one side of the street and stone on the other side. A 1-3-6 Portland cement concrete was used, a very rich mixture for a 6-in. foundation. The organization of the working gang was as follows:

Men loading and wheeling gravel	8
Men assisting in loading gravel	2
Man dumping barrows into hopper	1
Men loading and wheeling sand	3
Man dumping barrows into hopper	1
Men wheeling concrete in barrows	7
Men spreading concrete	3
Men tamping concrete	2
Man pouring cement into hopper	1
Man operating mixer	1
Man shoveling spilled concrete	1
Man opening cement bags	1
Engineer	1
Total men in gang	32

The average day's output of this gang was 150 cu. yds., or 900 sq. yds. in 8 hours; but on the best day's work the output was 200 cu. yds., or 1,200 sq. yds. in 8 hours, which is a remarkable record for 32 men and a mixer working only 8 hours.

The following is the labor cost of 8,896 sq. yds. of 4½-in. concrete foundation for an asphalt pavement constructed in New York City in 1904:

Item.	Per sq yd.

Foreman at \$3.75	\$0.030
Laborers at \$1.50	0.242
Teams at \$5	0.040
Steam engine at \$3.50	0.028
Total	\$0.340

The concrete was a 1-3-6 mixture and was mixed in a Foote mixer. These costs are compiled from data collected by the authors.

FOUNDATION CONSTRUCTION FOR STREET RAILWAY TRACK USING CONTINUOUS MIXERS.—The following account of the methods and cost of constructing a concrete foundation for street railway track at St. Louis, Mo., is compiled from information published by Mr. Richard McCulloch. The work was done by day labor by the United Railways Co., in 1906. Figure 114 shows the concrete construction. A 1-2½-6½ Portland cement, broken stone concrete mixed by machine was used.

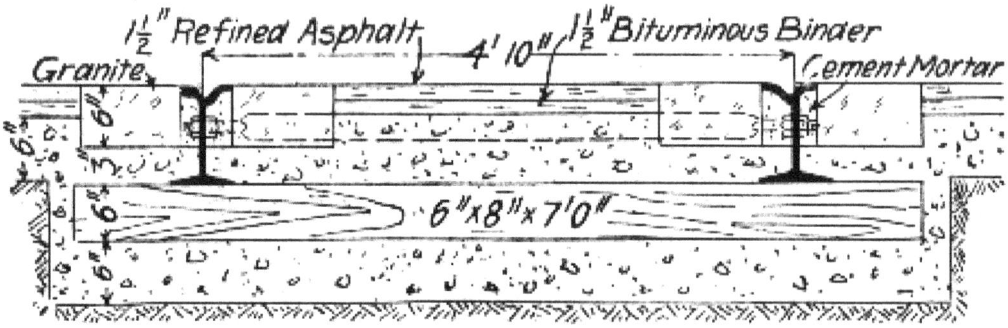

Fig. 114.—Concrete Foundation for Street Railway Track.

The material for the concrete was distributed on the street beside the tracks in advance of the machine, the sand being first deposited, then the crushed rock piled on that, and finally the cement sacks emptied on top of this pile. The materials were shoveled from this pile into the concrete mixing machine without any attempt at hand mixing on the street. Great care was taken in the delivery of materials on the street to have exactly the proper quantity of sand, rock and cement, so that there would be enough for the ballasting of the track to the proper height and that none would be left over. Each car was marked with its capacity in cubic feet, and each receiver was furnished with a table by which he could easily estimate the number of lineal feet of track over which the load should be distributed.

The concrete mixing machines were designed and built in the shops of the United Railways Co. Three machines were used in this work, one for each gang. The machine is composed of a Drake continuous worm mixer, fed by a chain dragging in a cast-iron trough. The trough is 36 ft. long, so that there is room for 14 men to shovel into it. Water is sprayed into the worm after the materials are mixed dry. This water was obtained from the fire plugs along the route. In the first machine built, the Drake mixer was 8 ft. long. In the two newer machines the mixer was 10 ft. long. Both the conveyor and the mixer were motor driven, current being obtained for this purpose from the trolley wire overhead. Two types of machines were used, one in which the conveyor trough was straight and 45 in. above the rail, and the other in which the conveyor trough was lowered back of the mixer, being 25 in. above the rail. The latter type had the advantage of not requiring such a lift in shoveling, but the trough is so low that a motor truck cannot be placed underneath it. In the high machine the mixer is moved forward by a standard motor truck under the conveyor. In the low machine the mixer is moved by a ratchet and gear on the truck underneath the mixer. A crew of 27 men is required to work each machine, and under average conditions concrete for 80 lin. ft. of single track, amounting to 22 cu. yds., can be discharged per hour.

The costs of the concrete materials delivered per cubic yard of concrete were: Cement, per barrel, $1.70; sand, per cu. yd., $0.675, and stone, per cu. yd., $0.425. The cost of the concrete work per cubic yard and per lineal foot of track was as follows:

Item.	Per lin. ft.	Per cu. yd.
Concrete materials	$0.791	$2.92
Labor mixing and placing	0.071	0.26
Total labor and materials	$0.862	$3.18

FOUNDATION CONSTRUCTION USING BATCH MIXERS AND WAGON HAULAGE, ST. LOUIS, MO.—The following record of the method and cost of laying a concrete foundation for street pavement using machine mixing and wagon haulage is given by Mr. D. A. Fisher. The foundation was 6 ins. thick. The gravel was dumped from wagons into a large hopper, raised by a bucket elevator into bins, and drawn off through gates into receiving hoppers on the charging platform where the cement was added. The receiving hoppers discharged into the mixers, which discharged the mixed concrete into a loading car that dumped into wagons, which delivered it on the street where wanted. The longest haul in wagons was 30 mins., but careful tests showed that the concrete had hardened well. The wagons were patent dump wagons of the drop-bottom type. Mr. Fisher says:

"You may consider the following figures a fair average of the plant referred to, working to its capacity. To these amounts, however, must be added the interest on the investment, the cost of wrecking the plant and the depreciation of the same, superintendence, and the pay roll that must be maintained in wet weather. I am assuming the street as already brought to grade and rolled.

"With labor at $1.75 per day of 10 hours, teams at $4, engineer and foremen at $3, and engine at $5 per day, concrete mixed and put in place by the above method costs:

	Per cu. yd.
To mix	$0.12 to $0.15
To deliver to street	0.10 to 0.14
To spread and tamp in place	0.08 to 0.11
Total	$0.30 to $0.40

"The mixers are No. 2½ Smith, sold by the Contractors' Supply Co., Chicago, Ill., and a ½ yd. cube, sold by Municipal Engineering & Contracting Co., Chicago.

"The above figures are on the basis of a batch every 2 minutes, which is easily maintained by using the loading car, as by this means there will be no delay in the operation of the plant owing to the irregularity of the arrival of the teams.

"My experience leads me to believe that a better efficiency can be obtained by using mixers of 1 cu. yd. capacity, and that the batch mixer is the only type of machine where any certainty of the proportion of the mixture is realized."

Fig. 115.—Chicago Improved Cube Traction Mixer for Pavement Foundation.

FOUNDATION CONSTRUCTION USING A TRACTION MIXER.— In laying a 6-in. foundation for an asphalt pavement in Buffalo, N. Y., an average of 100 sq. yds., or 16.6 cu. yds., of concrete in place was made per hour using the traction mixer shown by Fig. 115. This mixer was made by the Municipal Engineering & Contracting Co., of Chicago, Ill., and consisted of one of that company's improved cube mixers operated by a gasoline engine and equipped with the regulation mechanical charging device and also with a swinging conveyor to deliver the mixed concrete to the work. The feature of the apparatus in its application to paving work is the conveyor. This was 25 ft. long and pivoted at the mixer end so as to swing through an arc of 170°. The mixer discharged into a skip or bucket traveling on the conveyor frame and discharging over the end spreading its load anywhere within a radius of 25 ft. In operation the mixer traveled along the center of the street, backing away from the finished foundation and toward the stock pile, which was continuous and was deposited along the center of the street. The bulk of the sand and stone was thus shoveled direct into the charging bucket and the remainder was wheeled to the bucket in barrows. As the charging bucket is only 14 ins. high the barrows could be dumped directly into it from the ground. The gang worked was 17 including a foreman and one boy, and with this gang 100 sq. yds. of 6-in. foundation was laid per hour. Assuming an average wage of 20 cts. an hour the cost of mixing and placing the foundation concrete was 3.4 cts. per sq. yd. or 20.4 cts. per cu. yd. for labor alone.

FOUNDATION CONSTRUCTION USING CONTINUOUS MIXER. —The foundation was 6 ins. thick for an asphalt pavement and was laid in Chicago, Ill. The concrete used was exceptionally rich for pavement foundation work, it being a 1-3-6 Lehigh Portland cement, broken stone mixture. The mixing was done by machine, a mixer made by the Buffalo Concrete Mixer Co., Buffalo, N. Y., being used. This mixer was equipped with an elevating charging hopper and was operated as a continuous mixer. The mixer was mounted on wheels and was pulled along the center of the street ahead of the work with its discharge end toward the work. Moves of about 25 to 30 ft. were made, the mixer being pulled ahead for this distance each time that the concrete came up to its discharge end. The stock piles were continuous, sand on one side and stone on the other side of the street. Cement was stored in a pile at each end of the block. All materials were wheeled from stock piles to mixer in wheelbarrows. The men wheeling sand and stone loaded their own barrows, wheeled them to the mixer and discharged them directly into the elevating hopper. No runways were used, the barrows being wheeled directly on the ground. The cement was brought in barrows, two or three bags being a load, and dumped alongside a cement box which was located close to and at one side of the elevating hopper. A man untied the bags and emptied them into the cement box and another man scooped the cement out of the box in bucketfuls and emptied it over the sand and stone in the elevating hopper. The mixer discharged onto a sheet iron shoveling board, and the concrete was carried in shovels from shoveling board to place, the length of carry being a maximum of 25 to 30 ft. Two men were required to pull down the cone of concrete at the discharge end of the mixer and to keep the stone from separating and rolling down the sides. The gang was organized as follows:

	No. Men.
Loading and wheeling stone	10
Loading and wheeling sand	3 to 4
Loading and wheeling cement	2

Untieing and emptying cement bags	1
Charging cement to hopper	1
Operating mixer and hopper	1
Pulling down and tending discharge	2
Carrying concrete in shovels	8
Spreading concrete	2
Tamping concrete	2
Sweeping concrete	1
General laborers	3
Foreman	1
Watchman	1
Timekeeper	1
Total gang	40

This gang averaged 1,000 sq. yds. of 6-in. foundation per 10-hour day; a maximum of 1,400 sq. yds. was laid in a day. We have thus an average of 167 cu. yds. and a maximum of 234 cu. yds. of concrete foundation mixed and placed per 10-hour day. At an average wage of $2 per day the average labor cost of mixing and placing concrete was 48 cts. per cu. yd. or 8 cts. per sq. yd. of 6-in. foundation. It was stated that the gang was larger by three men than was ordinarily used owing to certain extra work being done at the time that the above figures were collected. Taking out three extra men and the timekeeper and watchman we get 34 men actually working in mixing and placing concrete. This reduced gang gives us a labor cost for mixing and placing of about 41 cts. per cu. yd. or 6.8 cts. per sq. yd. of 6-in. foundation.

FOUNDATION CONSTRUCTION USING A BATCH MIXER.—The following figures are an average of several jobs using a Ransome ½-cu. yd. mixer for constructing 6-in. foundations. The mixer was moved 1,000 ft. at a time and the work conducted 500 ft. in each direction from each station. The concrete materials were delivered from stock pile to mixer in wheelbarrows and the mixed concrete was hauled to the work in two-wheeled Ransome carts. Run planks were laid for the carts and one man readily pushed a cart holding 6 cu. ft. The men had to work fast on the long haul but had an easy time when the haul was short. The organization of the gang was as follows, wages being $1.50 per day:

10 men loading and wheeling stone	$15.00
4 men loading and wheeling sand	6.00
2 men handling cement	3.00
1 fireman	2.00
1 man dumping mixer	1.50
5 men wheeling carts	7.50
3 men spreading and ramming	4.50
1 foreman	3.50
Total wages per day	$43.00

This gang averaged 1,080 sq. yds. of 6-in. foundation or 180 cu. yds. of concrete in place per day which gives a labor cost of 24 cts per cu. yd. or 4 cts. per sq. yd. for mixing and placing.

Permissions

All chapters in this book are published with permission under the Creative Commons Attribution Share Alike License or equivalent. Every chapter published in this book has been scrutinized by our experts. Their significance has been extensively debated. The topics covered herein carry significant information for a comprehensive understanding. They may even be implemented as practical applications or may be referred to as a beginning point for further studies.

We would like to thank the editorial team for lending their expertise to make the book truly unique. They have played a crucial role in the development of this book. Without their invaluable contributions this book wouldn't have been possible. They have made vital efforts to compile up to date information on the varied aspects of this subject to make this book a valuable addition to the collection of many professionals and students.

This book was conceptualized with the vision of imparting up-to-date and integrated information in this field. To ensure the same, a matchless editorial board was set up. Every individual on the board went through rigorous rounds of assessment to prove their worth. After which they invested a large part of their time researching and compiling the most relevant data for our readers.

The editorial board has been involved in producing this book since its inception. They have spent rigorous hours researching and exploring the diverse topics which have resulted in the successful publishing of this book. They have passed on their knowledge of decades through this book. To expedite this challenging task, the publisher supported the team at every step. A small team of assistant editors was also appointed to further simplify the editing procedure and attain best results for the readers.

Apart from the editorial board, the designing team has also invested a significant amount of their time in understanding the subject and creating the most relevant covers. They scrutinized every image to scout for the most suitable representation of the subject and create an appropriate cover for the book.

The publishing team has been an ardent support to the editorial, designing and production team. Their endless efforts to recruit the best for this project, has resulted in the accomplishment of this book. They are a veteran in the field of academics and their pool of knowledge is as vast as their experience in printing. Their expertise and guidance has proved useful at every step. Their uncompromising quality standards have made this book an exceptional effort. Their encouragement from time to time has been an inspiration for everyone.

The publisher and the editorial board hope that this book will prove to be a valuable piece of knowledge for students, practitioners and scholars across the globe.

Index

www.ingramcontent.com/pod-product-compliance
Lightning Source LLC
LaVergne TN
LVHW080925210926
848485LV00004B/728